Math for Electricians

First Edition

Rob Zachariason

Publisher
The Goodheart-Willcox Company, Inc.
Tinley Park, IL
www.g-w.com

The Goodheart-Willcox Company, Inc. Brand Disclaimer: Brand names, company names, and illustrations for products and services included in this text are provided for educational purposes only and do not represent or imply endorsement or recommendation by the author or the publisher.

The Goodheart-Willcox Company, Inc. Safety Notice: The reader is expressly advised to carefully read, understand, and apply all safety precautions and warnings described in this book or that might also be indicated in undertaking the activities and exercises described herein to minimize risk of personal injury or injury to others. Common sense and good judgment should also be exercised and applied to help avoid all potential hazards. The reader should always refer to the appropriate manufacturer's technical information, directions, and recommendations; then proceed with care to follow specific equipment operating instructions. The reader should understand these notices and cautions are not exhaustive.

The publisher makes no warranty or representation whatsoever, either expressed or implied, including but not limited to equipment, procedures, and applications described or referred to herein, their quality, performance, merchantability, or fitness for a particular purpose. The publisher assumes no responsibility for any changes, errors, or omissions in this book. The publisher specifically disclaims any liability whatsoever, including any direct, indirect, incidental, consequential, special, or exemplary damages resulting, in whole or in part, from the reader's use or reliance upon the information, instructions, procedures, warnings, cautions, applications, or other matter contained in this book. The publisher assumes no responsibility for the activities of the reader.

The Goodheart-Willcox Company, Inc. Internet Disclaimer: The Internet resources and listings in this Goodheart-Willcox Publisher product are provided solely as a convenience to you. These resources and listings were reviewed at the time of publication to provide you with accurate, safe, and appropriate information. Goodheart-Willcox Publisher has no control over the referenced websites and, due to the dynamic nature of the Internet, is not responsible or liable for the content, products, or performance of links to other websites or resources. Goodheart-Willcox Publisher makes no representation, either expressed or implied, regarding the content of these websites, and such references do not constitute an endorsement or recommendation of the information or content presented. It is your responsibility to take all protective measures to guard against inappropriate content, viruses, or other destructive elements.

Image Credits. Front cover/Chapter opener: A_stockphoto/Shutterstock.com

Preface

Over the past 25 years that I have been teaching electrical courses, I have had many students become anxious when the subject of mathematics would come up. Many of the students who indicated they struggled with mathematics in the past were able to solve specific electrical problems when it was presented as a real-world situation. Relating math concepts to electrical scenarios helped with comprehension and ultimately allowed them to apply math principles to find the correct solution.

Math for Electricians looks at mathematics from a practical perspective. The text covers the basic math concepts used by electricians and other electrical industry professionals. *Math for Electricians* begins with a refresh of basic mathematical operations, such as addition, subtraction, multiplication, division, fractions, decimals, and percentages. The text continues with ratios, proportions, linear measurements, and conversions, and then covers algebraic formulas, geometry, and trigonometry. Each mathematical concept is related to an industry application with real-world examples and practice problems. The formulas and problems also align with subject matter covered in other electrical classes and can be applied to other trade areas. Chapter 10 accumulates all the mathematical concepts learned throughout the previous chapters and tests students' knowledge with a series of practical application questions. Highlighting where mathematics will be used on the job and relating it to other courses and real-world examples helps to understand its importance and engage students.

Rob Zachariason

About the Author

Rob Zachariason has worked in the electrical industry for over 30 years. He is currently an instructor in the Electrical Technology program at Minnesota State Community and Technical College in Moorhead, Minnesota. He was formerly an instructor for the Dakotas JATC in Fargo, North Dakota. When not in school, Mr. Zachariason is the working owner of Rob Zachariason Electric. Mr. Zachariason has also worked on several electrical textbooks, such as *Cracking the Code: A Practical Guide to the NEC*, as well as instructor supplements and videos. Mr. Zachariason has a diploma in Construction Electricity from Northwest Technical College and a bachelor's degree in Operations Management from Minnesota State University. He holds a Master Electrician license in both North Dakota and Minnesota and is a member of the International Brotherhood of Electrical Workers, the National Electrical Contractors Association, the International Association of Electrical Inspectors, the National Fire Protection Association, and the Minnesota State College Faculty union. Mr. Zachariason lives in Fargo with his wife, Brandi. They have three daughters, Lauren, Kate, and Julia.

Reviewers

The author and publisher wish to thank the following industry and teaching professionals for their valuable input into the development of *Math for Electricians.*

Neil Barker
Moraine Valley Community College
Palos Hills, Illinois

Andrea Blaylock
Triton College
River Grove, Illinois

Don Farrell
Centura College
Norfolk, Virginia

Brian D. Khairullah, MAE

Daniel Neff
Palm Beach State College
Lake Worth, Florida

Barbara Salazar
El Paso Community College
El Paso, Texas

Steve Senty
Anoka Technical College
Anoka, Minnesota

Matthew Wilkinson
Madison Area Technical College
Madison, Wisconsin

Features of the Textbook

The instructional design of this textbook includes student-focused learning tools to help you succeed. This visual guide highlights these features.

Chapter Opening Materials

Each chapter opener contains a list of learning objectives and a list of technical terms. **Objectives** clearly identify the knowledge and skills to be gained when the chapter is completed. **Technical Terms** list the key words to be learned in the chapter. **Introduction** provides an overview and preview of the chapter content.

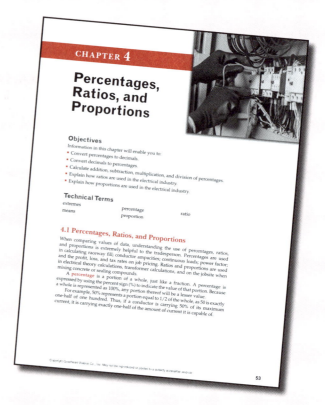

Additional Features

Additional features are used throughout the body of each chapter to further learning and knowledge. **Illustrations** have been designed to clearly and simply communicate the specific topic.

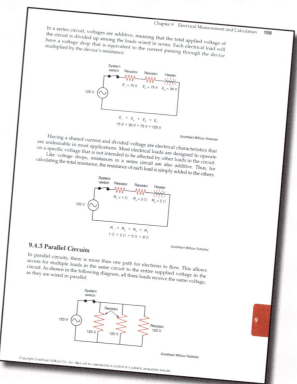

End-of-Chapter Content

End-of-chapter material provides an opportunity for review and application of concepts. **Review Questions** enable you to demonstrate knowledge, identification, and comprehension of chapter material. The questions also extend learning and develop your abilities to use learned material in new situations and to break down material into its component parts.

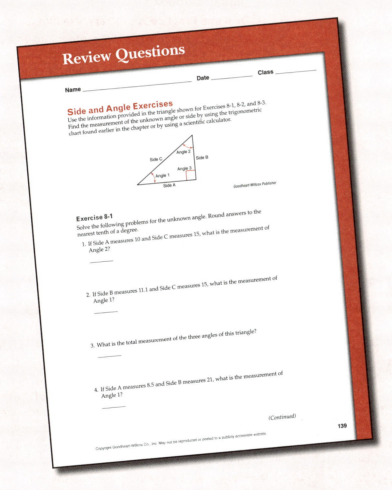

Review Questions

Name _____ Date _____ Class _____

Side and Angle Exercises

Use the information provided in the triangle shown for Exercises 8-1, 8-2, and 8-3. Find the measurement of the unknown angle or side by using the trigonometric chart found earlier in the chapter or by using a scientific calculator.

Goodheart-Willcox Publisher

Exercise 8-1

Solve the following problems for the unknown angle. Round answers to the nearest tenth of a degree.

1. If Side A measures 10 and Side C measures 15, what is the measurement of Angle 2?

   _____

2. If Side B measures 11.1 and Side C measures 15, what is the measurement of Angle 1?

   _____

3. What is the total measurement of the three angles of this triangle?

   _____

4. If Side A measures 8.5 and Side B measures 21, what is the measurement of Angle 1?

   _____

(Continued)

139

TOOLS FOR STUDENT AND INSTRUCTOR SUCCESS

Student Tools

Student Text

Math for Electricians is a write-in text that provides mathematical examples and exercises for an in-depth learning experience geared towards the electrical trade.

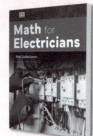

G-W Digital Companion

For digital users, e-flash cards and vocabulary exercises allow interaction with content to create opportunities to increase achievement.

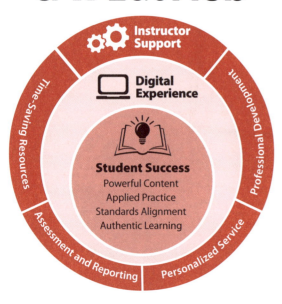

Instructor Tools

LMS Integration

Integrate Goodheart-Willcox content within your Learning Management System for a seamless user experience for both you and your students. EduHub LMS–ready content in Common Cartridge® format facilitates single sign-on integration and gives you control of student enrollment and data. With a Common Cartridge integration, you can access the LMS features and tools you are accustomed to using and G-W course resources in one convenient location—your LMS.

G-W Common Cartridge provides a complete learning package for you and your students. The included digital resources help your students remain engaged and learn effectively:

- **Digital Textbook**
- **Drill and Practice** vocabulary activities

When you incorporate G-W content into your courses via Common Cartridge, you have the flexibility to customize and structure the content to meet the educational needs of your students. You may also choose to add your own content to the course.

For instructors, the Common Cartridge includes the Online Instructor Resources. QTI® question banks are available within the Online Instructor Resources for import into your LMS. These prebuilt assessments help you measure student knowledge and track results in your LMS gradebook. Questions and tests can be customized to meet your assessment needs.

Online Instructor Resources

- The **Instructor Resources** provide instructors with time-saving preparation tools such as answer keys, editable lesson plans, and other teaching aids.
- **Instructor's Presentations for PowerPoint**® are fully customizable, richly illustrated slides that help you teach and visually reinforce the key concepts from each chapter.
- Administer and manage assessments to meet your classroom needs using **Assessment Software with Question Banks**, which include hundreds of matching, completion, multiple choice, and short answer questions to assess student knowledge of the content in each chapter.

See www.g-w.com/math-for-electricians-2025 for a list of all available resources.

Professional Development

- Expert content specialists
- Research-based pedagogy and instructional practices
- Options for virtual and in-person Professional Development

Brief Contents

Contents

CHAPTER 1

Whole Numbers

Objectives

Information in this chapter will enable you to:

- Identify how digits and place values are used in positive and negative numbers.
- Compare sets of whole positive and negative numbers and use the greater than, less than, and equals sign symbols.
- Apply rounding to numbers, up or down, as used in cost estimations.
- Calculate addition, subtraction, multiplication, and division of whole numbers.
- Determine the square and cube of a whole number.
- Express both large and small numbers using exponents, powers of ten, and scientific notation.
- Use operations with negative numbers.
- Calculate expressions using the PEMDAS order of operation.

Technical Terms

addition
borrowing
carried over
cube
difference
digit
dividend
division
division sign (÷)

divisor
equals sign (=)
exponent
factor
minus sign (−)
multiplication
multiplication
  sign (×)
negative

order of
  operation
place value
plus sign (+)
positive
powers of ten
product
quotient
rounding

scientific notation
square
subtraction
sum
superscript

1.1 Introduction to Whole Numbers

Whole numbers are the building blocks for the application of mathematics in everyday use. An electrician relies on this foundation to form equations and formulas for calculations, such as Ohm's law, box fill, and general lighting calculations, as well as everyday tasks like preparing an invoice to give to the customer. An understanding of the use of whole numbers, both positive and negative, will help in the further study of fractions, decimals, and formulas.

Whole numbers are formed by writing one or more digits in a row. A **digit** is any of the ten number symbols (0, 1, 2, 3, 4, 5, 6, 7, 8, and 9) in the Arabic numbering system. Each digit has a value, which increases from zero (0) through nine (9). A value of four (4), therefore, is greater than a value of one (1), but is less than a value of eight (8).

1.1.1 Positive and Negative Numbers

Numbers are **positive** if their value is greater than zero. Thus, the number 1 is positive, as it is greater than 0. Numbers are **negative** if their value is less than zero. Negative numbers are indicated by the minus sign (–) in front of the number. Therefore, the number –1 has a value below 0.

Electricians seldom work with negative numbers other than calculations involving low temperatures. Calculations concerning negative numbers have special rules, which will be covered in a later portion of this chapter.

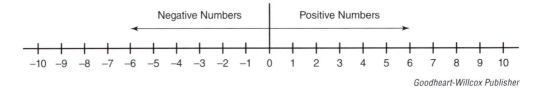

Goodheart-Willcox Publisher

1.1.2 Place Value of Whole Numbers

The **place value** of each digit is determined by its location within the whole number. The first ten place names in this numbering system from lower value to higher value are ones, tens, hundreds, thousands, ten thousands, hundred thousands, millions, ten millions, hundred millions, and billions.

A digit in the tens place has a value of ten times that of the same digit in the ones place, and a digit in the hundreds place has a value of one hundred times that of the same digit in the ones place, and so on.

The whole number one billion (written as 1,000,000,000) has place values as shown. It has zeros (0) in all of the places except the billions place, where it has a one (1).

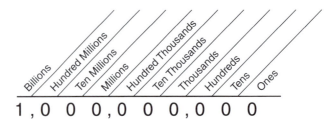

Goodheart-Willcox Publisher

The whole number three hundred seventy-five (375) has place values as shown. It has a three in the hundreds place, a seven in the tens place, and a five in the ones place.

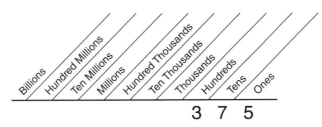

3 7 5

The value of the whole number 375 is explained as shown.

$$
\begin{aligned}
3 \times 100 &= 300 \\
7 \times 10 &= 70 \\
5 \times 1 &= \underline{+\ \ \ 5} \\
&\ \ \ \ 375
\end{aligned}
$$

The use of commas to separate the place values is commonly used. A comma is placed between each three digits, starting at the right and working to the left. For example, one thousand may be written either as 1000 or 1,000.

1.1.3 Comparing Whole Numbers

It is often necessary to compare the value of whole numbers to see which one is larger or smaller. Counting the number of digits is the easiest way to compare, as the number with more digits is greater. If the numbers being compared have the same number of digits, then each place value is compared, working from left to right.

Symbols that mean *greater than* and *less than* are used when comparing numbers. The symbol > means *greater than*, and < means *less than*. The symbol =, called the *equals sign*, means the numbers are equal, or exactly the same.

Comparison Symbols

>	Greater than
<	Less than
=	Equal

For example, when comparing the numbers 251 and 75, the number of digits shows which number is larger. The number 251 has three digits, while 75 has only two digits. Thus, 251 is larger than 75. This can be written as 251>75, which reads as "251 is greater than 75." It could also be written 75<251, which reads as "75 is less than 251."

When two numbers each have the same number of digits, such as 175 and 251, the numbers are compared digit by digit, starting from the left. Since the 1 in 175 is less than the 2 in 251, 175<251.

If 275 and 251 are compared, the 2 in both numbers is equal, so the next digit is compared. Since 7 is greater than 5, the result is 275>251.

1.1.4 Rounding Whole Numbers

Rounding of whole numbers is used for estimating purposes, such as preparing a rough cost of a project based on the sum of the materials and labor costs.

When numbers are rounded, they can be *rounded up*, *rounded down*, or rounded to the nearest place value, depending on the reason for the estimation. For example, when estimating materials for an electrical job, the numbers are often rounded up to make sure there are enough materials to complete the project. The same may be true when estimating the cost of a project. However, an estimate will be more accurate if the numbers are rounded to the nearest place value.

Rounding is the process of increasing or decreasing the value of a number to the next digit. Whether a digit's value is increased or not is usually based on the value of the digit to the right. Therefore, when rounding the *ones place to the nearest ten*, the value of the *tens place* will either remain the same (rounding down) or will increase by one (rounding up). When rounding the *tens place to the nearest hundred*, the *hundreds place*, will either remain the same (rounding down) or will increase by one (rounding up). This is the same for all positive numbers.

The digits 1 to 4 are rounded down, while the digits 5 to 9 are rounded up. Thus, the whole number 53 can be rounded down to 50, while the whole number 57 would be rounded up to 60. In this example, the ones place digit was rounded to the tens place.

When rounding a number that has three or more digits, the final value can range from lower to higher based on which digit is being rounded. For example, the number 572 can be rounded down to 570 when rounding the ones place to the nearest ten. When rounding the tens place to the nearest hundred, 572 would be rounded up to 600. This one number (572) can be rounded down to 570 or up to 600. These rounded potentials differ by 30. Note that the digit to which a number is rounded will be based on the accuracy needed for the value.

1.2 Operations Using Whole Numbers

Working with numbers generally involves operations, such as addition, subtraction, multiplication, and division. The **equals sign (=)** is used to denote the answer to the calculation. When used in a column of numbers, the equals sign is represented by a line beneath the bottom number in the column, indicating that the column needs to be totaled.

1.2.1 Addition of Whole Numbers

Addition is the process of combining number values to find the **sum** of those values. Addition is indicated by the use of the **plus sign (+)**. When whole numbers are added, each place column is totaled starting at the ones column. For example, $1 + 3 = 4$ would be read as "one plus three equals four" and could also be written as:

$$\begin{array}{r} 1 \\ + \, 3 \\ \hline 4 \end{array}$$

If the total of the ones column exceeds ten, the one from the ten is **carried over** to the tens column. When carrying over digits in addition, it is often helpful to write the carried-over number at the top of the proper column. For example, when adding seven plus three, the one from the sum of 7 + 3 is carried over to the tens column and can be written at the top of the tens column as shown below.

$$\begin{array}{r} {}^{1}7 \\ +\ 3 \\ \hline 10 \end{array}$$

In addition of larger numbers, the process of carrying over continues from right to left as each column of numbers is totaled.

$$\begin{array}{r} 748 \\ +\ 284 \\ \hline ? \end{array}$$

Starting at the right, in the ones column, eight plus four equals twelve. The 2 from the ones column of the number twelve is placed below the equals bar, and the 1 from the tens column of the number twelve is carried over and placed at the top of the tens column.

$$\begin{array}{r} {}^{1}\ \\ 748 \\ +\ 284 \\ \hline 2 \end{array}$$

The tens column is then added. One plus four plus eight equals thirteen, so the 3 is placed below the equals bar, and the 1 is carried over and placed at the top of the hundreds column.

$$\begin{array}{r} {}^{1\ 1}\ \\ 748 \\ +\ 284 \\ \hline 32 \end{array}$$

The hundreds column is then added. One plus seven plus two equals ten, so the 0 is placed below the equals bar, and the 1 is carried over and placed at the top of the thousands column.

$$\begin{array}{r} {}^{1\ 1}\ \\ {}^{1}748 \\ +\ 284 \\ \hline 032 \end{array}$$

The thousands column is then added. The only digit in that column is the 1 that was carried over from the hundreds column, so it is placed below the equals bar.

$$\begin{array}{r} {}^{1\ 1}\ \\ {}^{1}748 \\ +\ 284 \\ \hline 1,032 \end{array}$$

Regardless of the number of columns, all addition functions are done in this manner, working from right to left.

1.2.2 Subtraction of Whole Numbers

Subtraction is the process of removing, or taking away, number values from one another to find the difference between those values. In an equation, subtraction is indicated by the use of the **minus sign (–)**. When whole numbers are subtracted, each place column is subtracted individually, starting at the ones column and working from right to left.

For example, 4 – 3 = 1 would be read as "four minus three equals one" or "four take away three equals one" and could also be written as:

$$
\begin{array}{r}
4 \\
-\ 3 \\
\hline
1 \ \text{Difference}
\end{array}
$$

The **difference** is the answer to a subtraction equation.

When performing a subtraction operation with numbers greater than ten, first subtract the numbers in the ones place.

$$
\begin{array}{r}
67 \\
-\ 25 \\
\hline
2
\end{array}
$$

Then subtract the numbers in the tens place.

$$
\begin{array}{r}
67 \\
-\ 25 \\
\hline
42
\end{array}
$$

Sometimes the value of the digit in the number being subtracted is greater than the value of the digit in the number being subtracted from. When this happens, you "borrow" from the next-higher place value by adding 10 to the digit being subtracted from and reducing the value of the next-higher place value by 1. This is a process called **borrowing**.

$$
\begin{array}{r}
14 \\
-\ 7 \\
\hline
\end{array}
\qquad
\begin{array}{r}
^{0}\!\!\not{1}4 \\
-\ 7 \\
\hline
7
\end{array}
$$

In the example shown above, seven needs to be subtracted from four. Since seven is larger than four, the 1 from the tens column is *borrowed*, thus becoming fourteen minus seven, rather than four minus seven. As the value is borrowed, it is removed from that column, and crossing it out helps to clarify the operation. In this case, the 1 was borrowed, leaving 0. Seven is then subtracted from fourteen, and the difference (7) is written below the equals bar.

Borrowing can continue throughout the subtraction process, if necessary, with each column able to borrow from the columns to the left.

$$
\begin{array}{r}
747 \\
-\ 254 \\
\hline
\end{array}
$$

To begin the subtraction process in this example, four is subtracted from seven, leaving three, which is written below the equals bar.

$$
\begin{array}{r}
747 \\
-\ 254 \\
\hline
3
\end{array}
$$

Moving one column to the left (the tens column), five is then to be subtracted from four. However, since five is larger than four, a 1 (with a value of 10) is borrowed from the hundreds column, making the four into fourteen. To do this, the seven in the hundreds column is crossed out, leaving six remaining in the hundreds column. Fourteen minus five is nine, so 9 is written in the tens column.

$$
\begin{array}{r}
^{6}\!\!\not{7}47 \\
-\ 254 \\
\hline
93
\end{array}
$$

The hundreds column can now be subtracted. Six minus two equals four, so 4 is written in the hundreds column.

$$
\begin{array}{r}
\overset{6}{\cancel{7}}47 \\
-\ 254 \\
\hline
493
\end{array}
$$

Subtraction can easily be checked by addition. To check the answer in the example shown, the subtracted number (254) is added to the difference (493). The sum of these two numbers should equal the number that was subtracted from (747).

$$
\begin{array}{r}
\overset{1}{2}54 \\
+\ 493 \\
\hline
747
\end{array}
$$

Electricians use subtraction on a regular basis. A few common examples of when subtraction is used are determining how much material was used on a job, locating bend marks when cutting and bending conduit, and determining spacing and distances when installing boxes and lights.

When the material is delivered to a job, the quantity delivered will be documented. After the work has been completed, the remaining material will be returned to the shop. The amount of material returned is subtracted from the amount delivered to determine how much was used.

Quantity delivered – Quantity returned = Quantity used

100 switches delivered – 54 switches returned = 46 switches used

1.2.3 Multiplication of Whole Numbers

Multiplication, like addition, is the process of combining number values to get a total. This could be accomplished by simply adding the numbers together repeatedly, but a faster method is to *multiply*. The numbers being multiplied are called **factors**, and the result of the multiplication is called the **product**.

The **multiplication sign** ($\times$) is the multiplication symbol used in arithmetic functions, while in formulas the dot ($\cdot$) may be used to indicate multiplication. This avoids confusion when the letter x is used in algebraic equations to denote an unknown quantity.

$$
\begin{array}{r}
\text{First factor} \\
\times\ \text{Second factor} \\
\hline
\text{Product}
\end{array}
$$

A multiplication table can be used to assist in making multiplication calculations. To use the table shown here, find one factor in the first column (along the left side) and the second factor in the top row. Note that the multiplication of any number by zero (0) results in an answer of zero (0).

Multiplication Table

	0	1	2	3	4	5	6	7	8	9	10	11	12	13	14	15
0	0	0	0	0	0	0	0	0	0	0	0	0	0	0	0	0
1	0	1	2	3	4	5	6	7	8	9	10	11	12	13	14	15
2	0	2	4	6	8	10	12	14	16	18	20	22	24	26	28	30
3	0	3	6	9	12	15	18	21	24	27	30	33	36	39	42	45
4	0	4	8	12	16	20	24	28	32	36	40	44	48	52	56	60
5	0	5	10	15	20	25	30	35	40	45	50	55	60	65	70	75
6	0	6	12	18	24	30	36	42	48	54	60	66	72	78	84	90
7	0	7	14	21	28	35	42	49	56	63	70	77	84	91	98	105
8	0	8	16	24	32	40	48	56	64	72	80	88	96	104	112	120
9	0	9	18	27	36	45	54	63	72	81	90	99	108	117	126	135
10	0	10	20	30	40	50	60	70	80	90	100	110	120	130	140	150
11	0	11	22	33	44	55	66	77	88	99	110	121	132	143	154	165
12	0	12	24	36	48	60	72	84	96	108	120	132	144	156	168	180
13	0	13	26	39	52	65	78	91	104	117	130	143	156	169	182	195
14	0	14	28	42	56	70	84	98	112	126	140	154	168	182	196	210
15	0	15	30	45	60	75	90	105	120	135	150	165	180	195	210	225

Goodheart-Willcox Publisher

To calculate 2 × 3, which would be read as "two times three," locate the first factor (2) in the first column. Follow that row of numbers to the right.

Find the second factor (3) in the top row. Follow that column of numbers down. At the intersection of the two factors, a six (6) is shown. That is the product of 2 × 3.

When you are performing a multiplication operation, the order of the factors does not change the resulting product. That is, both 2 × 3 and 3 × 2 are equal to 6. (You can see this in the multiplication table.)

Multiplying two single-digit numbers (that is, 0–9) is normally accomplished by memorizing the multiplication table. Many techniques can be used to multiply larger numbers. The technique described in the following paragraph is one commonly used procedure.

In multiplication of whole numbers with more than one digit, each digit in the second factor is multiplied by each digit in the first factor, starting in the ones column and moving from right to left. For example, when multiplying 12 × 3, the 2 is multiplied by 3, which equals 6. The 6 is placed below the equals bar in the ones column.

$$\begin{array}{r} 12 \\ \times\ 3 \\ \hline 6 \end{array}$$

The 1 is then multiplied by 3. Because the 1 is in the tens column, it is actually 3 times 10, or 30. However, only 3 is placed below the equals bar in the tens column.

$$\begin{array}{r} 12 \\ \times\ 3 \\ \hline 36 \end{array}$$

Thus, $3 \times 12 = 36$.

If the multiplication of the numbers results in an answer of more than one digit, the tens place digit is carried over to the next column to the left. For example, when multiplying 25×3, the 5 is multiplied by 3, which equals 15. The 5 is written below the equals bar in the ones column, and the 1 is carried over to the top of the tens column.

$$\begin{array}{r} {\scriptstyle 1} \\ 25 \\ \times\ 3 \\ \hline 5 \end{array}$$

The 2 is then multiplied by 3, which is 6. The 1 that was carried over is added to the 6, making 7. The 7 is placed below the equals bar in the tens column.

$$\begin{array}{r} {\scriptstyle 1} \\ 25 \\ \times\ 3 \\ \hline 75 \end{array}$$

This process is repeated for larger numbers, always working from right to left.

If there is more than one digit in the second factor, the process described previously is repeated for each digit.

For example, in the multiplication of 25×43, the first step is to multiply 5 times 3, which equals 15. The 5 is written below the equals bar in the ones column, and the 1 is carried over to the top of the tens column.

$$\begin{array}{r} {\scriptstyle 1} \\ 25 \\ \times\ 43 \\ \hline 5 \end{array}$$

The 2 is then multiplied by 3, which is 6. The 1 that was carried over is added to the 6, making 7. The 7 is written below the equals bar in the tens column.

$$\begin{array}{r} {\scriptstyle 1} \\ 25 \\ \times\ 43 \\ \hline 75 \end{array}$$

The next step is to work with the multiplier digit in the tens column, which is the 4. Before starting that, realize that the 4 is in the tens column. Therefore, it is multiplying by 40 rather than 4. With this in mind, a zero is written in the ones place of the product of this multiplication.

$$\begin{array}{r} {\scriptstyle 1} \\ 25 \\ \times\ 43 \\ \hline 75 \end{array}$$
0 (Placeholder)

After the placeholder zero is added, begin multiplying using the digit in the tens column. The 4 is multiplied by the farthest right-hand digit in the first factor, which is in the ones column. Since 4×5 equals 20, the 0 is placed below the equals bar in the tens column, and the 2 is carried over to the top of the hundreds column.

$$
\begin{array}{r}
\overset{\overset{2}{\cancel{1}}}{25} \\
\times\ 43 \\
\hline
75 \\
00
\end{array}
$$

The 2 in the tens place is then multiplied by the 4, which equals 8. The 2 that was carried over and placed at the top of the hundreds column is added to the 8, making 10. The 0 is placed below the equals bar in the hundreds column, and the 1 is carried over to the top of the thousands column.

$$
\begin{array}{r}
\overset{\overset{2}{\cancel{1}}}{{}^{1}25} \\
\times\ 43 \\
\hline
75 \\
000
\end{array}
$$

Because all of the multiplier digits have been used, the 1 in the thousands column is simply brought down below the equals bar in the thousands column.

$$
\begin{array}{r}
\overset{\overset{2}{\cancel{1}}}{{}^{1}25} \\
\times\ 43 \\
\hline
75 \\
1000
\end{array}
$$

Once each digit of the second factor has been multiplied by all of the digits in the first factor, another equals bar is drawn, and the two products are added together, working from right to left.

$$
\begin{array}{r}
\overset{\overset{2}{\cancel{1}}}{{}^{1}25} \\
\times\ 43 \\
\hline
75 \\
+\ 1000 \\
\hline
1{,}075
\end{array}
$$

Thus, $25 \times 43 = 1{,}075$.

1.2.4 Squares, Cubes, and Exponents

The **square** of a whole number is that number multiplied by itself. A square is written using the exponent 2, which is shown as a **superscript**, placed after and above the whole number. An **exponent** is a number or symbol denoting the power to which another number is to be raised. For example, the square of 5 is written as 5^2 and read as "five squared." It is also called "five to the second power."

$$5^2 = 5 \times 5 = 25$$

The **cube** of a whole number is that number multiplied by itself twice. A cube is written using the exponent 3. The cube of 5 is written as 5^3 and read as "five cubed" or "five to the third power."

$$5^3 = 5 \times 5 \times 5 = 125$$

While exponents are seldom used by service technicians in the field, it is important to know because they might be found in equipment specifications, formulas, or engineering data.

1.2.5 Scientific Notation

In the engineering and mathematical sciences, **scientific notation** is often used as a simpler form of writing numbers that are too large or too small to be conveniently written in a standard format. Scientific notation uses powers of ten to express numerical values. While technicians rarely encounter numbers written in this form, a basic understanding of scientific notation will be helpful.

Powers of ten use an exponent to denote to which power of ten a number should be multiplied. With the exponent shown as a positive number, 10^2 means 10×10.

The number 3,000,000 (three million) is written in scientific notation as 3×10^6 and read as "three times ten to the sixth power." It means $3 \times 10 \times 10 \times 10 \times 10 \times 10 \times 10$. Essentially, the exponent of a positive power of ten indicates how many zeros belong to the right of the number raised to the power of ten.

Powers of Ten

Power	Power of Ten	Example
First power	10^1	$3 \times 10^1 = 30$
Second power	10^2	$3 \times 10^2 = 300$
Third power	10^3	$3 \times 10^3 = 3,000$
Fourth power	10^4	$3 \times 10^4 = 30,000$
Fifth power	10^5	$3 \times 10^5 = 300,000$
Sixth power	10^6	$3 \times 10^6 = 3,000,000$

Goodheart-Willcox Publisher

Thus, a number written as 3×10^3 means $3 \times 1,000$, or 3,000.

1.2.6 Division of Whole Numbers

Division is the process of separating numbers into groups of smaller numbers and is often thought of as the opposite of multiplication. If you have memorized the multiplication table, division becomes much easier to complete.

In division, the **dividend** is the number to be divided by the **divisor**, which results in the **quotient**. The **division sign (÷)** may be used when writing numbers to be divided, but common practice also includes simply placing the dividend over the divisor with a straight line between.

$$\text{Dividend} \div \text{Divisor} = \text{Quotient}$$

$$\frac{\text{Dividend}}{\text{Divisor}} = \text{Quotient}$$

To help achieve the division process, it is common to write the equation in a manner that allows easy calculation. The dividend is written below a calculation line, while the divisor is written to the left of the dividend. The quotient is then written on the calculation line.

$$\text{Divisor} \overline{\smash{)}\text{Dividend}}^{\text{Quotient}}$$

To divide 8 by 2, for example, the equation would be written as shown.

$$2\overline{\smash{)}8}$$

Since it is known that $2 \times 4 = 8$, then $8 \div 2 = 4$.

$$2\overline{\smash{)}8}^{\,4}$$

The 4 is placed on the top of the calculation line directly above the 8. The 4 is then multiplied by the 2 (the divisor), and the product of that multiplication (8) is placed below the dividend 8.

$$
\begin{array}{r}
4 \\
2\overline{\smash{)}8} \\
8
\end{array}
$$

An equals bar is placed below the product 8, and it is subtracted from the dividend 8.

$$
\begin{array}{r}
4 \\
2\overline{\smash{)}8} \\
-8 \\
\hline
0
\end{array}
$$

Because the difference is zero (0), the calculation is complete.

To check a quotient, use multiplication. Multiply the quotient by the divisor. The answer should equal the dividend. If it does, the quotient is correct. In this example, multiply 4 (quotient) times 2 (divisor) to get 8 (dividend).

If the value of the divisor does not equally go into the dividend, a remainder will result. For example, in the problem $8 \div 3$, the number 8 cannot be evenly divided by 3.

$$3\overline{\smash{)}8}$$

The number 3 will go into 8 twice, so 2 is written on the calculation line directly above the 8.

$$3\overline{\smash{)}8}^{\,2}$$

The quotient is then multiplied by the divisor (3×2), and the product (6) is written directly below the 8.

$$
\begin{array}{r}
2 \\
3\overline{\smash{)}8} \\
-6 \\
\hline
2
\end{array}
$$

The 6 is subtracted from the 8, leaving 2. If left in this form, the answer would be 2, with a remainder of 2. To check the accuracy of the division, multiply the quotient (2) by the divisor (3) and then add the remainder (2).

$$\text{Dividend} = (\text{Quotient} \times \text{Divisor}) + \text{Remainder}$$
$$= (2 \times 3) + 2$$
$$= 6 + 2$$
$$= 8$$

The answer is correct, as the 8 in the checking exercise matches the 8 in the dividend.

The use of remainders when doing division is seldom used other than in the process of learning how to divide whole numbers. Instead of leaving the remainder, the division process is carried beyond the whole number by the use of fractions or decimal places. Fractions and decimals will be covered in detail in later chapters of this text.

1.2.7 Operations Involving Negative Numbers

Negative numbers are numbers with a value less than zero. These are seldom encountered by people in the electrical industry. The most applicable electrical applications for negative numbers are when dealing with temperatures below zero degrees (0°) and calculating temperature differences.

The following table lists the rules to follow for basic math operations involving negative numbers:

Math Rules of Operation for Negative Numbers

Operation	Rules	Examples
Addition	If signs are the same, add the values and keep the sign.	3 + 4 = 7 (both positive) −3 + (−5) = −8 (both negative)
	If signs are different, subtract the values and keep the sign of the larger value.	8 + (−3) = 5 (positive value is larger) 7 + (−9) = −2 (negative value is larger)
Subtraction	Convert to an addition expression by (1) changing the minus sign to a plus sign and (2) changing the sign of the number being subtracted. Then follow the rules for addition.	13 − 7 becomes 13 + (−7) = 6 12 − (−4) becomes 12 + 4 = 16 −3 − 6 becomes −3 + (−6) = −9 −7 − (−12) becomes −7 + 12 = 5
Multiplication and Division	If signs are the same, the product or quotient is positive.	4 × 5 = 20 (signs are the same, product is positive) −3 × −8 = 24 (signs are the same, product is positive) 72 ÷ 9 = 8 (signs are the same, quotient is positive) −81 ÷ −9 = 9 (signs are the same, quotient is positive)
	If signs are different, the product or quotient is negative.	7 × −4 = −28 (signs are different, product is negative) −6 × 8 = −48 (signs are different, product is negative) 56 ÷ −7 = −8 (signs are different, quotient is negative) −50 ÷ 5 = 10 (signs are different, quotient is negative)

Goodheart-Willcox Publisher

When calculating a temperature difference, you always subtract the lesser (or colder) temperature from the greater (or warmer) temperature. Thinking of these values as warmer and colder may be helpful when temperatures include negative values.

For example, to determine the temperature difference between –12°F and 4°F, subtract the lesser (colder) temperature from the greater (warmer) temperature:

$$\text{Temperature difference} = \text{warmer temp} - \text{colder temp}$$
$$= 4°F - (-12°F)$$
$$= 4°F + 12°F$$
$$= 16°F$$

As a second example, to determine the temperature difference between –5°F and –17°F, subtract the lesser (colder) temperature from the greater (warmer) temperature:

$$\text{Temperature difference} = \text{warmer temp} - \text{colder temp}$$
$$= -5°F - (-17°F)$$
$$= -5°F + 17°F$$
$$= 12°F$$

When installing a raceway in an area where the temperature is going to fluctuate, such as on the outside of a building, consideration must be taken to how much the length of the raceway will change due to expansion and contraction. All raceways will expand and contract when exposed to changes in temperature, but polyvinyl chloride (PVC) is impacted more than metallic raceways. Expansion fittings will be installed to accommodate the variation in overall length of the conduit if it varies by more than 1/4 inch. *Table 352.44(A)* of the *National Electrical Code (NEC)* is used to determine the length the conduit will change due to variations in temperature. Before using the *Table*, you have to determine the temperature difference the raceway will be exposed to.

For example, a building with a PVC raceway on the exterior will be subjected to a winter low temperature of –12°F and a summer high temperature of 109°F. The temperature difference is calculated as follows:

$$\text{Temperature difference} = \text{warmer temp} - \text{colder temp}$$
$$= 109°F - (-12°F)$$
$$= 109°F + 12°F$$
$$= 121°F$$

After finding a temperature difference of 121°F, you can now go to *Table 352.44(A)* of the *NEC* and find the length of change of the PVC conduit. For 121°F, you would round up to the next highest temperature, 125°F, resulting in a length of change of 5.07 in/100 ft.

1.2.8 Combined Operations with Whole Numbers

It is often necessary to work with groups of whole numbers in a variety of operations; you may need to multiply, add, and subtract in order to accomplish the desired goal.

For example, calculating the work, travel, and break times of a service electricians day requires addition (the time spent on travel and lunch), subtraction (the total time in the workday minus the hours spent on travel and lunch), and division (the time spent on service divided by the number of service calls completed).

When completing a calculation involving multiple operations, the **order of operation** is critical to finding the correct answer. The acronym **PEMDAS** can be used to help remember the proper order.

Parentheses—Any operations inside parentheses are completed first, working from the innermost parentheses to the outermost.

Exponents—Any numbers containing exponents are completed second.

Multiplication/**D**ivision—Multiplication and division are completed in left to right order.

Addition/**S**ubtraction—Addition and subtraction are completed in left to right order.

NFPA 70®, National Electrical Code®, and NEC® are registered trademarks of the National Fire Protection Association, Quincy, MA.

Name _____   Date _____   Class _____

Whole Number Exercises

Exercise 1-1

Use the number 397,481 to answer the following questions.

1. Which digit is in the tens place?

   _____

2. Which digit is in the ten thousands place?

   _____

3. Which digit is in the ones place?

   _____

4. Which digit is in the thousands place?

   _____

Use the number 819,542 to answer the following questions.

5. In which place value is the 5?

   _____

6. In which place value is the 8?

   _____

7. In which place value is the 4?

   _____

8. In which place value is the 1?

   _____

Exercise 1-2

1. If a whole number has a 4 in the ten thousands place (its highest place value), what is the maximum value of that whole number?

   _____

2. What is the minimum value?

   _____

3. If a whole number has a 7 in the hundred thousands place (its highest place value), what is the maximum value of that whole number?

   _____

4. What is the minimum number?

   _____

Practical Exercise 1-3

An industrial electric boiler has a power rating (in watts) stated on the nameplate. The number shown has a seven (7) in the hundred thousands place, a five (5) in the hundreds place, a five (5) in the ten thousands place, a zero (0) in the ones place, a one (1) in the millions place, a zero (0) in the thousands place, and a zero (0) in the tens place.

    1. What is the power rating of the boiler?

       _____watts

Practical Exercise 1-4

An electric furnace needs service. As the electrician tries to read the nameplate for data, the numbers for the power rating of the unit are unreadable. There is room for 5 digits in the space that shows the watt rating.

    1. What is the maximum rating in watts this unit could be?

       _____watts

    2. What is the minimum rating in watts this unit could be?

       _____watts

Comparing Exercise

Exercise 1-5

Compare each pair of numbers and place the symbol <, >, or = in the space provided.

    1. 25 ____ 112

    2. 117 ____ 134

    3. 4,275 ____ 4,257

    4. 972 ____ 792

    5. 19,357 ____ 19,361

    6. 71 ____ 71

    7. 39,754 ____ 37,954

    8. 125 ____ 152

Rounding Exercise

Practical Exercise 1-6

The installation of lights in a warehouse will have metal clad (MC) cable running between the lights. The drawings provided by the design engineer indicate that 930 feet of MC cable will be needed.

    1. Round the amount of pipe to the nearest hundred.

_____

Name _____   **Date** _____   **Class** _____

2. The MC cable will be purchased in 250-foot rolls. Does this affect the manner in which the number is rounded? Explain the reasons why.

_____

_____

_____

Addition Exercises

Exercise 1-7

Add the following columns of numbers, showing all work.

1. $\begin{array}{r} 24 \\ + 13 \\ \hline \end{array}$

2. $\begin{array}{r} 35 \\ + 62 \\ \hline \end{array}$

3. $\begin{array}{r} 74 \\ + 51 \\ \hline \end{array}$

4. $\begin{array}{r} 44 \\ + 73 \\ \hline \end{array}$

5. $\begin{array}{r} 24 \\ + 67 \\ \hline \end{array}$

6. $\begin{array}{r} 48 \\ + 25 \\ \hline \end{array}$

7. $\begin{array}{r} 49 \\ + 73 \\ \hline \end{array}$

8. $\begin{array}{r} 57 \\ + 79 \\ \hline \end{array}$

9. $\begin{array}{r} 475 \\ + 13 \\ \hline \end{array}$

10. $\begin{array}{r} 438 \\ + 89 \\ \hline \end{array}$

11. $\begin{array}{r} 374 \\ + 189 \\ \hline \end{array}$

12. $\begin{array}{r} 4,355 \\ + 8,626 \\ \hline \end{array}$

13. $\begin{array}{r} 127 \\ 234 \\ + 398 \\ \hline \end{array}$

14. $\begin{array}{r} 3,412 \\ 12 \\ 987 \\ + 8,941 \\ \hline \end{array}$

15. $\begin{array}{r} 9,571 \\ 15,293 \\ + 31 \\ \hline \end{array}$

16. $\begin{array}{r} 17 \\ 287,492 \\ 345 \\ 9,150 \\ + 3 \\ \hline \end{array}$

17. $\begin{array}{r} 31 \\ 1,267 \\ 2 \\ + 99,681 \\ \hline \end{array}$

18. $\begin{array}{r} 56,020 \\ 234 \\ + 190,902 \\ \hline \end{array}$

19. $\begin{array}{r} 9,201 \\ 47 \\ 186 \\ + 2,220 \\ \hline \end{array}$

20. $\begin{array}{r} 996,712 \\ 561,390 \\ 121 \\ 9,845,621 \\ + 762,314 \\ \hline \end{array}$

Practical Exercise 1-8

A service electrician is asked to plan for the replacement of emergency light batteries throughout a large building complex. Each building has several emergency lights, which all take the same battery. Building A needs 17 batteries. Building B needs 22 batteries. Building C needs 12 batteries. Building D needs 32 batteries. Building E needs 114 batteries.

1. How many batteries should the electrician plan on purchasing for this job?

Practical Exercise 1-9

At the end of the workweek, each service electrician is asked to report the number of miles driven in their service vehicle for that week. A daily record is kept to show the miles driven each day. The record shows Monday at 187 miles, Tuesday at 112 miles, Wednesday at 49 miles, Thursday at 241 miles, and Friday at 99 miles.

1. What is the total mileage reported for the week?

Subtraction Exercises

Exercise 1-10

Subtract the following numbers, showing all work. Check all answers by using addition.

1. 42
   − 31

2. 329
   − 49

3. 9,476
   − 7,693

4. 3,487
   − 999

5. 97,567
   − 9,688

6. 178,245
   − 9,468

7. 3,875
   − 2,986

8. 769
   − 671

Practical Exercise 1-11

On an electric service upgrade, an electrician notices the customer's invoice shows a duplicate charge for a new meter socket assembly that was installed on the home. Thus, the subtotal on the bill is incorrect. The invoice shows a subtotal of $1,763, but the meter socket assembly was listed twice, each time for $597.

1. What is the correct subtotal?

Name _____ **Date** _____ **Class** _____

Practical Exercise 1-12

When wiring a new house, 2,500 feet of 12-2 nonmetallic sheathed cable is delivered by the parts runner. When the job is finished, 110 feet of the cable is returned to the shop.

1. How many feet of 12-2 nonmetallic sheathed cable was used when wiring the home?

_____

Combined Exercises

Exercise 1-13

Multiply the following numbers, showing all work.

1. 27
    $\times\ 4$

2. 141
    $\times\ 13$

3. 792
    $\times\ 24$

4. 23
    $\times\ 131$

5. 379
    $\times\ 158$

6. 1,316
    $\times\ 290$

7. 79,351
    $\times\ 21,083$

8. 90,048
    $\times\ 209$

9. $18^2 =$ _____

10. $12^3 =$ _____

11. $10^5 =$ _____

12. $7 \times 10^4 =$ _____

Practical Exercise 1-14

While a service electrician was replacing a faulty switch at a customer's home, the owner asked what it would cost to upgrade the six exterior lights. The electrician determined that it would take about 3 hours to replace all the lights. The electrical contractor charges $92 per hour for labor, and the material cost is $135 for each light.

1. What should the electrician estimate the total cost for the customer to be?

$ _____

Practical Exercise 1-15

An electric service company has 34 regular service vehicles on the road to do installation and repair work throughout a metropolitan area. Each vehicle travels an average of 100 miles per day, 5 days per week. Only 12 of the vehicles are used for weekend emergency service calls, averaging 140 miles on Saturday and 80 miles on Sunday.

1. What is the total mileage of all of the regular vehicles on a weekday?

   _____ miles

2. What is the total mileage of all of the regular vehicles in one week?

   _____ miles

3. What is the total mileage of all of the regular vehicles over a 4-week period?

   _____ miles

4. What is the total mileage of all of the emergency vehicles on a single Saturday?

   _____ miles

5. What is the total mileage of all of the emergency vehicles on a single Sunday?

   _____ miles

6. What is the total mileage of all of the emergency vehicles over a 4-week period?

   _____ miles

7. How many total miles do all of the vehicles travel in 4 weeks?

   _____ miles

Division Exercises

Exercise 1-16

Complete the following division problems. Write the quotient as a whole number and, when applicable, a remainder. Check all answers by using multiplication.

1. $40 \div 8 =$ _____

2. $126 \div 14 =$ _____

3. $24 \div 5 =$ _____

4. $2{,}180 \div 109 =$ _____

5. $1{,}428 \div 9 =$ _____

6. $96 \div 3 =$ _____

Name _____   **Date** _____   **Class** _____

7. 12,568 ÷ 8 = _____

8. 23,902 ÷ 105 = _____

9. 89,460 ÷ 21,348 = _____

Practical Exercise 1-17

Over one week's time, a service electrician worked 54 hours and made 18 service calls.

1. If each service call lasted the same amount of time, how long did each service call last? Include the travel time between calls with each service call.

_____

Negative Number Exercise
Practical Exercise 1-18

Rigid PVC conduit is going to be installed on the outside of a building from the meter socket to the disconnect location. In this geographic location, the low temperature in the winter could reach −28°F, and the high temperature in the summer could reach 106°F.

1. What is the temperature difference that the raceway will be exposed to?

_____

Combined Operations Exercises

Solve each equation, showing your work.

Practical Exercise 1-19

An electric service company has 32 service vehicles on the road to do installation and repair work throughout the area. Each vehicle travels 90 miles per day, 5 days per week. The vehicles average 12 miles per gallon of gas, and the price of gas is $2.95 per gallon.

1. How many gallons of gas will the company have to purchase each week? The problem would be set up as shown.

$$\frac{(32 \times 90 \times 5)}{12} = \_\_\_\_\_$$

2. How much will the company spend per week on fuel? The problem would be set up as shown.

$$\frac{(32 \times 90 \times 5)}{12} \times 2.95 = \_\_\_\_\_$$

(Continued)

3. What is the cost per vehicle of fuel each week? Round the answer to the nearest cent (two decimal places). The problem would be set up as shown.

$$\frac{(32 \times 90 \times 5)}{12} \times 2.95 \div 32 = \underline{\hspace{1cm}}$$

Practical Exercise 1-20

An office building has 23 rooftop air-handling units (AHUs) to heat and cool the building. Each AHU holds 4 filters, costing $7.90 each, that need to be changed every 90 days. It takes a maintenance worker 1/2 hour to change the filters in each AHU, at a rate of $72.00 per hour.

1. What is the cost of the filters for all of the AHUs every 90 days?

   _____

2. What is the cost of the labor to change the filters every 90 days?

   _____

3. What is the total cost per day of keeping the filters changed?

   _____

CHAPTER 2

Fractions

Objectives

Information in this chapter will enable you to:

- Define each part of a fraction.
- Identify proper fractions, improper fractions, and mixed numbers.
- Compare, reduce, and raise fractions.
- Calculate the lowest common denominator (LCD) among fractions.
- List the factors of numbers and determine the greatest common factor.
- Identify the most common prime numbers.
- Calculate fractions to be reduced to their lowest terms.
- Calculate addition, subtraction, multiplication, and division of fractions.
- Convert a mixed number into an improper fraction.

Technical Terms

common
  denominator
denominator
factor

fraction
greatest common
  factor
improper fraction

lowest common
  denominator
  (LCD)
lowest terms
mixed number

numerator
prime number
proper fraction
reducing

2.1 Introduction to Fractions

A **fraction** is defined as a portion of a whole. A fraction, therefore, is part of a whole number. Electricians commonly work in situations with fractions, such as conduit and tubing sizing (1/2 inch, 3/4 inch, 1 1/4 inch, etc.) and using fractions of an inch on tape measures. An understanding of fractions, their use, and basic mathematical functions, such as adding and subtracting fractions, is critical in the trade.

If the number 1 is broken into four equal portions, each portion would be 1/4 of the whole.

Four Equal Portions of a Whole

Goodheart-Willcox Publisher

In a fraction, the top number, the number above the line, is called the **numerator**. The bottom number, the number below the line, is called the **denominator**.

$$\frac{\text{Numerator}}{\text{Denominator}} \quad \frac{1}{4}$$

The line dividing the numerator and denominator denotes division. Therefore, the fraction 1/4 means 1 divided by 4.

Fractions are commonly used in measurements, as the unit of measurement may not be accurate enough to be in whole number form. For example, when measuring a length of conduit using feet as the unit of measurement, it may not be accurate enough to find that the pipe is over 7 feet long but under 8 feet long. Therefore, for the sake of accuracy, a smaller unit of measurement is used. The inch, which is a fraction of a foot, would be more accurate. Because there are 12 inches in 1 foot, 1 inch is a fraction equivalent to 1/12 of a foot.

Linear Measurement – US Customary Units

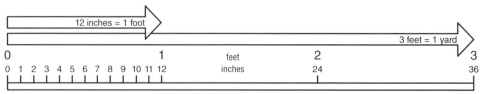

Goodheart-Willcox Publisher

If the inch as a unit of measurement is not accurate enough to suit the purpose, a fraction of an inch may be used. Commonly used fractions for linear measurement in the electrical industry go as small as 1/16 of an inch. Smaller fractions will be encountered for items such as drill bits, but accuracy to the nearest 1/16th of an inch is sufficient for most situations.

Fractions of an Inch

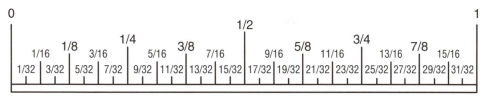

Goodheart-Willcox Publisher

Various units of measurement are fully discussed in Chapter 5, *Linear Measurements and Conversions.*

2.1.1 Proper and Improper Fractions

A **proper fraction** is defined as a fraction in which the numerator is a smaller number than the denominator. The fraction 1/4 is an example of a proper fraction.

Proper Fractions

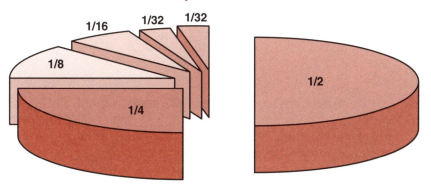

Goodheart-Willcox Publisher

An **improper fraction** is defined as a fraction in which the numerator is a larger number than the denominator. The fraction 3/2 would be considered an improper fraction. Improper fractions are commonly changed to mixed numbers.

2.1.2 Mixed Numbers

A **mixed number** consists of a whole number followed by a proper fraction. The improper fraction 3/2 can be changed to a mixed number by division. Since the fraction 3/2 means 3 ÷ 2, it can be calculated as:

$$2\overline{)3}$$

Since 2 will go into 3 one time, the 1 is written above the 3 on the calculation line.

$$2\overline{)3}^{\,1}$$

Since 1 times 2 is 2, the 2 is subtracted from the 3, leaving 1.

$$\begin{array}{r} 1 \\ 2\overline{)3} \\ -2 \\ \hline 1 \end{array}$$

Since 2 cannot be divided into 1, the remainder 1 can be written as a proper fraction as 1 over 2. The 1/2 is added to the whole number 1 (the quotient), making the final answer 1 1/2. In this way, the improper fraction 3/2 has now been changed to the mixed number 1 1/2.

$$\frac{3}{2} = 1\frac{1}{2}$$

2.1.3 Comparing, Reducing, and Raising Fractions

In fractions, the higher the number in the denominator, the more parts of a whole there are. Therefore, the higher the number of the denominator, the smaller each individual part of the numerator. For example, 1/8 is less than 1/4, which is less than 1/2.

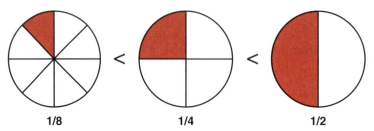

When comparing fractions to see which has the higher value, both the numerator and denominator must be compared. Compare 1/4 and 1/2. Each fraction has the same number in the numerator. Therefore, the value of the denominator will determine which fraction has the greater value.

$$\frac{1}{4} < \frac{1}{2}$$

However, not all fractions have the same number in their numerators. For example, compare 3/4 and 1/2. Even though the denominator 4 is a greater number than the denominator 2, the fraction 3/4 is not less than 1/2. Comparing denominators to determine which fraction is greater can only be done when the numerators are equal.

To compare fractions with different numerators, their denominators must be the same. Once each fraction has the same denominator, then the values of their numerators can be compared. To do this, a **common denominator** must be found.

The simplest way to find a common denominator is to multiply the denominators by each other, resulting in a denominator that is common to both. For example, to compare 3/4 and 1/2, multiply the denominators of the fractions (4 × 2). This results in the finding of a common denominator of 8.

Numerators can be raised or lowered as long as the ratio of the numerator to the denominator remains the same. To maintain this ratio, the numerator and denominator of a fraction are each multiplied by the same number.

$$\frac{3}{4} \times \frac{2}{2} = \frac{6}{8}$$

On the top, 3 was multiplied by 2. On the bottom, 4 was multiplied by 2. Since ratios were maintained, the value of the fraction has not changed. Therefore, according to this example:

$$\frac{3}{4} = \frac{6}{8}$$

Similarly, 1/2 can be raised to have a denominator of 8. This will allow the values of the fractions 3/4 (now 6/8) and 1/2 to be compared. To raise the denominator of 1/2 to 8, multiply the fraction by 4/4. To maintain ratios, multiply the denominator by 4 and the numerator by 4.

$$\frac{1}{2} \times \frac{4}{4} = \frac{4}{8}$$

Since the denominator was multiplied by 4, the numerator was also multiplied by 4. This keeps the ratio between the numerator and denominator the same.

$$\frac{1}{2} = \frac{4}{8}$$

Now that the two fractions have a common denominator, they can be easily compared by looking at the value of the numerators. Since the numerator 6 is larger than the numerator 4, 6/8 (3/4) is larger than 4/8 (1/2). This could be written as:

$$\frac{3}{4} > \frac{1}{2}$$

When comparing fractions, the size of the common denominator is not important. However, when solving math problems involving fractions, it is helpful to use the **lowest common denominator (LCD)** (also called the *least common denominator*) in order to simplify the calculation.

In order to determine the lowest common denominator (LCD), several methods may be employed. However, the simplest way is to compare the factors of each denominator, looking for the **greatest common factor**. A **factor** is defined as an integer that divides evenly into another integer. In common language, a factor is a whole number that divides into another whole number.

For example, determine the factors of the number 6. Since 6 ÷ 1 = 6, both 1 and 6 are factors. Since 6 ÷ 3 = 2, both 2 and 3 are factors. No other whole numbers can be used to divide into 6. Therefore, the numbers 1, 2, 3, and 6 are factors of 6.

The best factors are prime numbers. A **prime number** is a number that has only two factors: 1 and itself. The prime numbers most commonly used as factors are 2, 3, 5, and 7. Many numbers are not prime numbers. For example, 4 is not a prime number because 4 has three factors: 1, 2, and 4. To quickly determine the greatest common factor of fractions, become acquainted with the more commonly used prime numbers.

Lowest Prime Numbers

2	3	5	7
11	13	17	19
23	29	31	37
41	43	47	53
59	61	67	71

Goodheart-Willcox Publisher

In the previous example of comparing fractions, 8 is a common denominator, but it is not the lowest common denominator. Below are the factors of each denominator used in that example.

For 3/4:

$$1 \times 4 \ = \ 4$$
$$2 \times 2 \ = \ 4$$

Therefore, factors of 4 are 1, 2, and 4.

For 1/2:

$$1 \times 2 \ = \ 2$$

Therefore, factors of 2 are 1 and 2.

From the list of the factors shown for each denominator, the common factors are 1 and 2. Since 2 is the higher of those numbers, 2 is the greatest common factor of the numbers 2 and 4. By multiplying the two denominators by each other and then dividing that number by the greatest common factor, the lowest common denominator is easily found.

$$\text{Denominator} \times \text{Denominator} = \text{Common denominator}$$

$$\frac{\text{Common denominator}}{\text{Greatest common factor}} = \text{Lowest common denominator (LCD)}$$

Thus,

$$\text{Common denominator} = 4 \times 2$$
$$= 8$$

$$\text{LCD} = \frac{\text{Common denominator}}{\text{Greatest common factor}}$$

$$= \frac{8}{2}$$

$$\text{Lowest common denominator} = 4$$

If the fractions to be compared are 9/32 and 11/18, multiplying 32 times 18 results in a common denominator of 576. To find a lower common denominator, look at the factors of each denominator.

For 9/32:

$$1 \times 32 \ = \ 32$$
$$2 \times 16 \ = \ 32$$
$$4 \times 8 \ = \ 32$$

Therefore, factors of 32 are 1, 2, 4, 8, and 32.

For 11/18:

$$1 \times 18 \ = \ 18$$
$$2 \times 9 \ = \ 18$$
$$3 \times 6 \ = \ 18$$

Therefore, factors of 18 are 1, 2, 3, 6, 9, and 18.

The only common factors of 32 and 18 are 1 and 2. The greatest common factor of 32 and 18 is 2. To find the lowest common denominator when 32 and 18 are the denominators, multiply the denominators together, and then divide by the greatest common factor (2).

$$\text{Common denominator} = 32 \times 18$$

$$= 576$$

$$\text{LCD} = \frac{\text{Common denominator}}{\text{Greatest common factor}}$$

$$= \frac{576}{2}$$

$$= 288$$

The LCD of 32 and 18 is 288. The fractions of 9/32 and 11/18 would then be converted to the LCD of 288. To determine which number to multiply with each numerator, divide 288 by each denominator.

$$288 \div 32 \ = \ 9$$

Now multiply the numerator by 9 and the denominator by 9.

$$\frac{9}{32} \times \frac{9}{9} = \frac{81}{288}$$
$$288 \div 18 = 16$$

Now multiply the numerator by 16 and the denominator by 16.

$$\frac{11}{18} \times \frac{16}{16} = \frac{176}{288}$$

The fractions 9/32 and 11/18 can now be compared. Because 81/288 is less than 176/288, 9/32 is less than 11/18. This can be written as:

$$\frac{9}{32} < \frac{11}{18}$$

Fractions can be raised or reduced, as long as the ratio between the numerator and the denominator stays the same. In comparing fractions, raising the fraction is quite often necessary in order to compare their values.

Calculations involving fractions are easiest to work with if they are reduced to their **lowest terms. Reducing** a fraction simply means lowering both the numerator and the denominator to the lowest possible number, while maintaining the ratio between the two. This can be done by seeing if the same prime number is divisible into both the numerator and the denominator of the fraction to be reduced.

For example, the fraction 4/8 can be reduced. Start with the prime number 2 and check to see if both the numerator (4) and the denominator (8) are evenly divisible by 2.

$$\frac{4 \div 2}{8 \div 2} = \frac{2}{4}$$

Since both the numerator and the denominator are evenly divided by the same number, the fraction 4/8 is reduced to 2/4. To see if further reduction is possible, try dividing both the numerator and denominator again by the highest possible prime number that divides evenly into both. Since 2 is again the highest prime number that can be used in this situation, divide both by 2.

$$\frac{2 \div 2}{4 \div 2} = \frac{1}{2}$$

The fraction 4/8 has now been reduced to 1/2. The fraction 1/2 cannot be reduced further, as the only number that divides evenly into both is 1. In this way, 4/8 has been reduced to its lowest terms.

2.2 Operations Using Fractions

Just like whole numbers, fractions can be added, subtracted, multiplied, and divided. However, mathematical operations involving fractions must follow certain rules.

2.2.1 Addition and Subtraction of Fractions

Fractions can be added and subtracted as long as the denominators are the same. When the denominators are the same, the numerators can be added or subtracted as if they are whole numbers.

$$\frac{1}{8} + \frac{5}{8} = \frac{6}{8}$$

In some cases, adding and subtracting fractions requires converting the fractions to a common denominator. Once a common denominator has been found and the fractions have the same denominator, the numerators can be added or subtracted.

For example, to add 1/4 and 1/2, the denominator must be converted to match. Since 1/2 is the same as 2/4, the calculation can be performed by raising the fraction 1/2 to 2/4.

$$\frac{1}{4} + \frac{1}{2} = \frac{1}{4} + \frac{2}{4}$$

When fractions are added or subtracted, only the numerators are added and subtracted. The denominator remains the same. Therefore, when adding 1/4 and 2/4, 1 plus 2 equals 3. The numerator of the sum is 3, and the denominator remains 4. The sum of 1/4 and 2/4 (1/2) is 3/4.

$$\frac{1}{4} + \frac{2}{4} = \frac{3}{4}$$

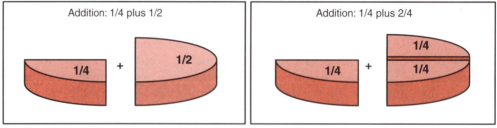

Goodheart-Willcox Publisher

If the fraction 1/4 is subtracted from 1/2, the denominator needs to be the same before the numerators can be subtracted. Since 1/2 can be raised to 2/4, 1/4 can then be subtracted from 2/4 by simply working with the numerators.

$$\frac{1}{2} - \frac{1}{4} = \frac{2}{4} - \frac{1}{4}$$

$$\frac{2}{4} - \frac{1}{4} = \frac{1}{4}$$

As shown, 1/2 minus 1/4 equals 1/4.

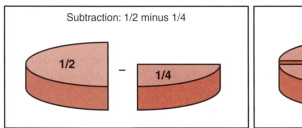

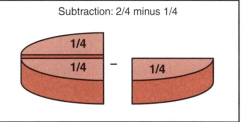

2.2.2 Multiplication of Fractions and Mixed Numbers

Multiplication of fractions is similar to multiplying whole numbers because the numerators and denominators are multiplied separately. Therefore, in multiplication of fractions, it is not necessary to find a common denominator. Once the multiplication is complete, the fraction is then reduced to its lowest possible terms.

For example, to multiply 1/2 times 1/2, it would be set up as shown.

$$\frac{1}{2} \times \frac{1}{2}$$

The numerators are multiplied together. Therefore, 1 times 1 equals 1, which is the numerator of the answer. The denominators are then multiplied by each other. Since 2 times 2 equals 4, the denominator of the answer is 4.

$$\frac{1}{2} \times \frac{1}{2} = \frac{1}{4}$$

With any mathematical problem, always make sure the answer makes sense. In this case, does it make sense that one-half times one-half equals one-fourth?

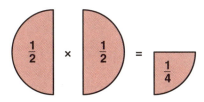

It may seem odd that $1/2 \times 1/2$ equals 1/4. After all, 1/4 is smaller than 1/2. Usually when multiplying numbers, the answer is larger than the two numbers being multiplied. However, this is different with fractions and decimals.

When performing multiplication, you can think of the product as the total quantity in a number of same-sized groups. For example, 4×3 can be thought of as counting the total quantity in three groups, with each group containing four items. When multiplying by a fraction, you are determining the quantity contained in a portion of the entire group. For example, $4 \times 1/2$ can be thought of as determining the quantity of one-half of the group.

In the earlier example, the equation $1/2 \times 1/2$ is asking, "What is the value of one-half of one-half?" Half of one-half (1/2) is one-quarter (1/4) because it would take two one-quarters to equal one-half (1/4 +1/4 = 2/4, reduced to 1/2). In other words, when multiplying by a proper fraction, the product will be less than the original value.

Multiplying mixed numbers by fractions or whole numbers requires the mixed number to be changed to an improper fraction. Whole numbers must also be converted to a fraction when multiplying with fractions.

To convert a mixed number to an improper fraction, multiply the whole number portion by the denominator of the fraction portion, and then add that product to the numerator. Continue using the original denominator.

$$\text{Mixed number} = (\text{Whole number} \times \text{Denominator}) + \frac{\text{Numerator}}{\text{Denominator}}$$

For example, to convert the mixed number 2 1/4 to an improper fraction, multiply the whole number (2) by the denominator (4). Add the product of that multiplication (8) to the numerator (1).

$$2\frac{1}{4} = \frac{(2 \times 4) + 1}{4}$$

$$= \frac{8 + 1}{4}$$

$$= \frac{9}{4}$$

As shown, 2 1/4 can be changed to the improper fraction 9/4, which can now be easily used in multiplication calculations.

Whole numbers must also be converted to fraction form when multiplying by fractions. Since a whole number is always a quotient of itself and 1, any whole number can be written as itself divided by 1.

$$\text{Whole number} = \frac{\text{Whole number}}{1}$$

For example, the number 7 can be written as 7 over 1 with no change in value.

$$\frac{7}{1}$$

Thus, to multiply 7 times 1/3:

$$\frac{7}{1} \times \frac{1}{3}$$

Multiplying the numerator times the numerator and the denominator times the denominator results in the improper fraction 7/3.

$$\frac{7}{1} \times \frac{1}{3} = \frac{7}{3}$$

The improper fraction is then reduced to a mixed number by dividing the numerator by the denominator.

$$\frac{7}{3} = 2\frac{1}{3}$$

2.2.3 Division of Fractions and Mixed Numbers

The division rule for fractions can simply be remembered as *invert and multiply*. To divide fractions, the first fraction (the dividend) is multiplied by the inverse of the second fraction (the divisor). To invert a fraction, simply switch the numerator and denominator.

$$\text{Dividend} \div \text{Divisor} = \text{Dividend} \times \text{Inverted Divisor}$$

For example, to divide 1/2 by 1/4, the problem would initially be set up as:

$$\frac{1}{2} \div \frac{1}{4} =$$

The second fraction is then inverted, and the two fractions are multiplied. The numerators are multiplied by each other, and the denominators are multiplied by each other.

$$\frac{1}{2} \div \frac{1}{4} = \frac{1}{2} \times \frac{4}{1} = \frac{4}{2}$$

In this example, the result is an improper fraction. This can then be reduced by dividing the numerator by the denominator, resulting in a whole-number answer. In many instances, the answer may be a mixed number.

$$\frac{1}{2} \div \frac{1}{4} = \frac{1}{2} \times \frac{4}{1}$$
$$= \frac{4}{2}$$
$$= 2$$

When dividing mixed numbers and fractions, the mixed number must first be converted to an improper fraction. For example, to divide 3 1/2 by 1/4, the problem would initially be set up as shown.

$$3\frac{1}{2} \div \frac{1}{4} =$$

The 3 1/2 is first converted to the improper fraction 7/2 by multiplying the denominator (2) times the whole number (3) and then adding the numerator (1): $(2 \times 3) + 1 = 7$. The denominator remains the same.

$$\frac{7}{2} \div \frac{1}{4} =$$

The second fraction is then inverted, and the fractions are multiplied: the numerator times the numerator and the denominator times the denominator.

$$\frac{7}{2} \times \frac{4}{1} = \frac{28}{2}$$

The improper fraction is then reduced by dividing the numerator by the denominator.

$$28 \div 2 = 14$$

Thus,

$$3\frac{1}{2} \div \frac{1}{4} = 14$$

Like mixed numbers, whole numbers must also be changed to fraction form in order to perform division with fractions. For example, to divide 2 by 1/4, the problem is initially set up as shown.

$$2 \div \frac{1}{4} =$$

The whole number 2 can be converted to a fraction by dividing the 2 by 1.

$$\frac{2}{1} \div \frac{1}{4} =$$

The second fraction (the divisor) is then inverted, and the two fractions are multiplied: the numerator times the numerator and the denominator times the denominator.

$$\frac{2}{1} \times \frac{4}{1} = \frac{8}{1}$$

Since 8 divided by 1 equals 8, the answer to 2 ÷ 1/4 is 8.

Review Questions

Name _____ **Date** _____ **Class** _____

Fractions Exercises

Exercise 2-1

Change each fraction or mixed number to its lowest possible form. Show all work.

1. $\dfrac{12}{16}$ = _____

2. $\dfrac{7}{8}$ = _____

3. $\dfrac{9}{12}$ = _____

4. $7\dfrac{1}{2}$ = _____

5. $\dfrac{18}{12}$ = _____

6. $\dfrac{4}{10}$ = _____

7. $\dfrac{60}{100}$ = _____

8. $\dfrac{27}{9}$ = _____

9. $\dfrac{5}{15}$ = _____

10. $\dfrac{8}{3}$ = _____

11. $\dfrac{7}{21}$ = _____

12. $9\dfrac{2}{7}$ = _____

Exercise 2-2

Reduce all fractions to their lowest value. Enter that value on the first blank. Next, determine the lowest common denominator of these fractions. Convert each fraction to have that denominator, and list that form of the fraction on the second blank.

1. $\dfrac{9}{12}$ = _____ , _____

2. $\dfrac{8}{12}$ = _____ , _____

3. $\dfrac{5}{16}$ = _____ , _____

4. $\dfrac{5}{8}$ = _____ , _____

5. $\dfrac{1}{16}$ = _____ , _____

6. $\dfrac{1}{32}$ = _____ , _____

7. $\dfrac{1}{2}$ = _____ , _____

8. $\dfrac{7}{8}$ = _____ , _____

9. $\dfrac{9}{16}$ = _____ , _____

10. $\dfrac{13}{16}$ = _____ , _____

11. $\dfrac{31}{32}$ = _____ , _____

12. $\dfrac{15}{32}$ = _____ , _____

13. $\dfrac{11}{16}$ = _____ , _____

14. $\dfrac{1}{3}$ = _____ , _____

(Continued)

15. $\frac{4}{4}$ = _____ , _____

16. $\frac{21}{32}$ = _____ , _____

17. $\frac{15}{16}$ = _____ , _____

18. $\frac{34}{64}$ = _____ , _____

Exercise 2-3

List the fractions from Exercise 2-2 in order from lowest to highest. In each entry, include the original fraction and its most reduced form.

1. _____ , _____

2. _____ , _____

3. _____ , _____

4. _____ , _____

5. _____ , _____

6. _____ , _____

7. _____ , _____

8. _____ , _____

9. _____ , _____

10. _____ , _____

11. _____ , _____

12. _____ , _____

13. _____ , _____

14. _____ , _____

15. _____ , _____

16. _____ , _____

17. _____ , _____

18. _____ , _____

Practical Exercise 2-4

You are installing a raceway between an electrical panel and a junction box. There is a leftover piece of conduit from another cut that is 46 5/8″ long. The distance between the panel and the box is 46 13/16″ long. Is the scrap piece of conduit long enough, or will you need to cut a new piece? Explain the reason for your answer and show all work.

_____

_____

_____

Addition and Subtraction Exercise

Exercise 2-5

Add or subtract the following fractions. Show all work, including finding the common denominator when necessary. Change any answers that are improper fractions into mixed numbers. Reduce fractions to the lowest terms.

1. $\frac{1}{2} + \frac{3}{4}$ = _____

2. $\frac{7}{8} - \frac{3}{16}$ = _____

3. $\frac{9}{10} - \frac{3}{5}$ = _____

4. $\frac{7}{16} + \frac{3}{8}$ = _____

Name _____ **Date** _____ **Class** _____

5. $\dfrac{9}{16} - \dfrac{1}{2} =$ _____

6. $1\dfrac{1}{2} - \dfrac{7}{8} =$ _____

7. $3\dfrac{3}{4} + 7\dfrac{1}{2} =$ _____

8. $14\dfrac{9}{16} - 12\dfrac{7}{8} =$ _____

9. $7\dfrac{7}{16} + 9\dfrac{5}{8} =$ _____

10. $\dfrac{8}{32} + \dfrac{1}{4} =$ _____

11. $3\dfrac{3}{16} + 7\dfrac{1}{2} =$ _____

12. $17\dfrac{3}{32} - 9\dfrac{1}{8} =$ _____

Multiplication Exercise

Exercise 2-6

Multiply the following fractions and mixed numbers. Reduce all answers to the lowest terms. Show all work.

1. $\dfrac{1}{2} \times \dfrac{7}{8} =$ _____

2. $\dfrac{1}{4} \times \dfrac{9}{16} =$ _____

3. $\dfrac{5}{6} \times \dfrac{1}{3} =$ _____

4. $\dfrac{1}{10} \times 3\dfrac{4}{5} =$ _____

5. $7\dfrac{3}{8} \times 12\dfrac{1}{2} =$ _____

6. $15 \times \dfrac{1}{4} =$ _____

7. $\dfrac{5}{8} \times \dfrac{5}{8} =$ _____

8. $\dfrac{6}{7} \times 1 =$ _____

9. $45\dfrac{1}{2} \times \dfrac{1}{4} =$ _____

10. $1\dfrac{7}{8} \times 2 =$ _____

11. $1\dfrac{1}{4} \times 1\dfrac{1}{4} =$ _____

12. $9 \times \dfrac{1}{3} =$ _____

Division Exercises

Exercise 2-7

Divide the following fractions and mixed numbers. Reduce all answers to the lowest terms. Show all work.

1. $\dfrac{1}{2} \div \dfrac{1}{2} =$ _____

2. $\dfrac{1}{2} \div \dfrac{1}{4} =$ _____

3. $1\dfrac{2}{3} \div \dfrac{3}{4} =$ _____

4. $3\dfrac{5}{8} \div 2 =$ _____

(Continued)

5. $\frac{7}{8} \div \frac{1}{2} =$ _____

9. $\frac{11}{16} \div 1 =$ _____

6. $10\frac{7}{16} \div 2\frac{5}{8} =$ _____

10. $\frac{15}{16} \div 1\frac{15}{16} =$ _____

7. $\frac{9}{12} \div 3\frac{5}{7} =$ _____

11. $\frac{3}{4} \div \frac{1}{2} =$ _____

8. $1\frac{21}{32} \div 2\frac{1}{3} =$ _____

12. $9\frac{7}{8} \div 3 =$ _____

Practical Exercise 2-8

An apprentice electrician is asked to cut conduit nipples for use on a project. Each piece needs to be exactly 4 3/4 inches long.

1. How many pieces can be cut from a 10-foot length of conduit?

_____

2. Assuming no mistakes are made in cutting, and ignoring the amount of conduit lost during the cutting process, what will be the length of conduit remaining after all pieces have been cut?

_____

Practical Exercise 2-9

A wall in a warehouse has an electrical panel and is to have 12 new receptacles installed. The distance to the first receptacle is 65 1/2 inches, and the remaining receptacles are spaced so there are 52 3/4 inches between them.

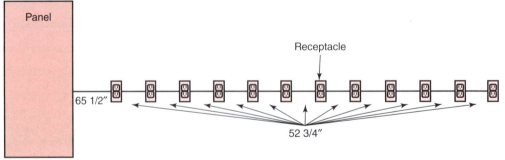

Goodheart-Willcox Publisher

1. Ignoring the loss from the cutting process, what is the total length of conduit required for this installation?

_____

2. The raceway is available in 10-foot (120-inch) lengths. If you want to complete the installation without using any couplings, how many lengths of conduit will you need?

_____

CHAPTER 3

Decimals

Objectives

Information in this chapter will enable you to:

- Convert fractions to decimals.
- Convert decimals to fractions.
- Calculate addition, subtraction, multiplication, and division of decimals.

Technical Terms

decimal   decimal equivalent   decimal point

3.1 Introduction to Decimals

In the electrical industry, decimals are used as frequently as fractions. In some instances, decimals are used more often. For example, when doing calculations, an electronic calculator is commonly used. Since these instruments do not normally recognize entries such as 1/4, 1/2, or 3/4, these commonly used fractions must be converted to decimal form (for example, 0.25, 0.5, and 0.75) to be entered into a calculator.

In many situations encountered in the trades, manufacturers will supply equipment data using decimals, such as a motor rated at 17.25 horsepower (hp).

A **decimal** is a fractional portion (less than one) of a number expressed using place values to the right of a decimal point. These decimal place values represent denominators of powers of ten, such as tenths, hundredths, and thousandths.

The position of a **decimal point** in relation to each digit of a number determines the assigned value. This is similar to the digits in whole numbers. The decimal point delineates the value of the digit. Digits to the left of the decimal point have a value greater than 1, while digits to the right of the decimal point have a value of less than 1.

3.1.1 Place Value

Whole numbers are generally written without a decimal point at the right of the number. Adding a decimal point will not change their value. The place value of each digit is based on its location in the number. For digits to the left of a decimal point, the farther left, the greater the value of the digit. For digits to the right of a decimal point, the farther right, the lesser the value of the digit.

Place Values

Hundreds	Tens	Ones		Tenths	Hundredths	Thousandths
1	2	3	.	4	5	6

Goodheart-Willcox Publisher

In this example of 123.456, the 1 is located in the greatest place value, because it is the farthest to the left of the decimal point. The 6 is located in the least place value, because it is farthest to the right of the decimal point.

Working from the right of the decimal point, the first four place names in this numbering system from greater value to lesser value are tenths, hundredths, thousandths, and ten thousandths. Decimal places can be smaller than ten thousandth, but these are not commonly used in the trade.

A digit in the tenths place has a value of one-tenth that of the same digit in the ones place, and a digit in the hundredths place has a value of one-hundredth that same digit in the ones place, and so on.

Any number can be written with or without a decimal point. If the value includes a fraction, however, the fraction can be written as a decimal. For example, the number 1 1/4 written in decimal form would be 1.25 as shown.

Place Values

Thousands	Hundreds	Tens	Ones		Tenths	Hundredths	Thousandths	Ten Thousandths
0	0	0	1	.	2	5	0	0

Goodheart-Willcox Publisher

3.1.2 Converting Fractions to Decimal Form

All fractions can be converted to decimal form by changing the fraction's denominator to 10 or a multiple of 10, such as 100 or 1,000. For example, if the fraction 1/4 is to be converted to decimal form, the denominator 4 must be converted to 10 or a multiple of 10.

$$\frac{1}{4} = \frac{?}{10}$$

The fraction bar or slash means divide, so 1/4 actually means 1 divided by 4.

$$4\overline{)1}$$

Since 4 will not go into 1, a decimal point is placed to the right of the 1, and a 0 is placed to the right of the decimal in the tenths place. The decimal is also added above the division line in the same place. To ensure that digits are not placed in the wrong location, a 0 is added to the left of the decimal point above the division line.

$$\frac{0.}{4\overline{)1.0}}$$

The divisor 4 can be divided into 10. Since 4 goes into 10 two times, a 2 is added above the line but to the right of the decimal point.

$$\frac{0.2}{4\overline{)1.0}}$$

The 2 placed to the right of the decimal point is the tenths place value, so the 2 is valued at two-tenths (2/10). Since 2×4 is 8, an 8 is then subtracted from 10, leaving 2.

$$\begin{array}{r} 0.2 \\ 4\overline{)1.0} \\ -8 \\ \hline 2 \end{array}$$

Because 4 will not divide into 2, another 0 is added to the dividend. This 0 is carried down to 20. The divisor 4 divides into 20 by 5. The 5 is placed to the right of the 0.2 in the hundredths place, making its value five hundredths (5/100).

$$\begin{array}{r} 0.25 \\ 4\overline{)1.00} \\ -8\downarrow \\ \hline 20 \end{array}$$

The 5 is then multiplied by 4, which equals 20. This value is subtracted from the 20 in the dividend and leaves no remainder.

$$\begin{array}{r} 0.25 \\ 4\overline{)1.00} \\ -8\downarrow \\ \hline 20 \\ -20 \\ \hline 0 \end{array}$$

Thus, the fraction 1/4 divides 1 by 4 to equal 0.25. In fraction form, this would be 25/100. It is read as "twenty-five hundredths" or "twenty-five one hundredths." This is called the **decimal equivalent** of a fraction.

It is common in the electrical industry to use decimals rather than fractions for calculations. All fractions can be converted to decimal form, and a decimal equivalent chart can be used to speed up the conversion process.

Common Decimal Equivalents

Fraction	Decimal	Fraction	Decimal	Fraction	Decimal	Fraction	Decimal
1/64	0.015625	17/64	0.265625	33/64	0.515625	49/64	0.765625
1/32	0.03125	9/32	0.28125	17/32	0.53125	25/32	0.78125
3/64	0.046875	19/64	0.296875	35/64	0.546875	51/64	0.796875
1/16	0.0625	5/16	0.3125	9/16	0.5625	13/16	0.8125
5/64	0.078125	21/64	0.328125	37/64	0.578125	53/64	0.828125
3/32	0.09375	11/32	0.34375	19/32	0.59375	27/32	0.84375
7/64	0.109375	23/64	0.359375	39/64	0.609375	55/64	0.859375
1/8	0.125	3/8	0.375	5/8	0.625	7/8	0.875
9/64	0.140625	25/64	0.390625	41/64	0.640625	57/64	0.890625
5/32	0.15625	13/32	0.40625	21/32	0.65625	29/32	0.90625
11/64	0.171875	27/64	0.421875	43/64	0.671875	59/64	0.921875
3/16	0.1875	7/16	0.4375	11/16	0.6875	15/16	0.9375
13/64	0.203125	29/64	0.453125	45/64	0.703125	61/64	0.953125
7/32	0.21875	15/32	0.46875	23/32	0.71875	31/32	0.96875
15/64	0.234375	31/64	0.484375	47/64	0.734375	63/64	0.984375
1/4	0.25	1/2	0.5	3/4	0.75	1	1.0

Goodheart-Willcox Publisher

The zero to the left of the decimal point is shown when there is no whole number in that position. The zero reminds us that the decimal is there, which otherwise could be overlooked. The number of places to the right of the decimal point determines the precision of the decimal equivalent. In many cases, decimals will be rounded off to a certain number of decimal places. In the electrical industry, rounding to the second or third decimal place is normal.

3.1.3 Converting Decimals to Fractions

To convert decimals to fractions, simply write any digits to the right of the decimal point as the numerator of the fraction. The denominator will be 1 followed by zeros. The number of zeros used in the denominator will be the same number as the number of digits in the numerator.

$$0.1 = \frac{1}{10}$$

$$0.01 = \frac{1}{100}$$

$$0.001 = \frac{1}{1,000}$$

After a decimal has been converted to a fraction, the fraction may be able to be reduced to lesser terms. For example, the decimal 0.2 can be converted to 2/10. This can be reduced to 1/5.

$$0.2 = \frac{2}{10}$$

$$\frac{2}{10} = \frac{1}{5}$$

Just as positive exponents are used to express large numbers, negative exponents can be used to express small numbers. For example, 10^{13} would equal 10,000,000,000,000. A negative exponent (such as 10^{-3}) means the reciprocal of $10 \times 10 \times 10$. A reciprocal is placed beneath a one and a fraction bar. It becomes one over that number.

$$10^{-3} = \frac{1}{10 \times 10 \times 10} = \frac{1}{1,000}$$

Therefore, if a number was written as 3×10^{-3}, it would mean:

$$3 \times \frac{1}{1,000} = \frac{3}{1,000} = 0.003$$

When a number is multiplied with a ten to a negative exponent (10^{-X}), the decimal point moves to the left. For a negative exponent, do not multiply the number by the 10. Simply move the decimal point to the left the number of times as the exponent.

$$3 \times 10^{-3} = 0.003$$

Calculations involving numbers written in scientific notation are done in the same manner as calculations involving any other whole number or decimal.

3.2 Mathematic Operations Involving Decimals

Operations using decimals are handled exactly the same as operations involving whole numbers. Addition, subtraction, multiplication, and division are all done in the same manner, but the position of the decimal point must be closely watched.

3.2.1 Addition of Decimals

When adding decimals, the decimal points must be aligned vertically, regardless of the number of digits on either side of the decimal point. For example, when adding 3.1 to 14.74, the operation would be arranged as follows.

$$\begin{array}{r} 3.1 \\ + 14.74 \\ \hline \end{array}$$

To avoid confusion, extra zeros (**0**) may be added as placeholders to the right of the last digit in a decimal. The numbers are then added just as in whole numbers, and the decimal point in the sum is in the same position vertically.

$$\begin{array}{r} 3.1\mathbf{0} \\ + 14.74 \\ \hline 17.84 \end{array}$$

3.2.2 Subtraction of Decimals

Just like addition, subtraction of decimals is handled exactly the same as the subtraction of whole numbers. The position of the decimal point will be aligned vertically. For example, to subtract 4.24 from 8.7936, the calculation would be set up as shown.

$$
\begin{array}{r}
8.7936 \\
-\,4.24 \\
\hline
\end{array}
$$

Extra zeros may be added as placeholders to avoid misalignment of the decimal point. The numbers are then subtracted just as if they were whole numbers, and borrowing may be done as necessary. The decimal point in the difference is aligned vertically with the decimal points above.

$$
\begin{array}{r}
8.7936 \\
-\,4.2400 \\
\hline
4.5536 \\
\end{array}
$$

3.2.3 Multiplication of Decimals

When multiplying decimals, the decimal point is not aligned vertically as it is in addition and subtraction. The numbers to be multiplied are written as if there were no decimal point, and the multiplication is then done the same as in multiplying whole numbers. For example, if multiplying 1.5×7, the calculation would be as follows:

$$
\begin{array}{r}
1.5 \\
\times\,7 \\
\hline
105 \\
\end{array}
$$

Once the multiplication has been completed, the position of the decimal point in each number is counted from right to left and added together. In the example used here, 1.5 has a decimal point that is one place from the right, and 7 has no decimal. Since $1 + 0 = 1$, the decimal point in the answer is placed one space from the right, making the final answer 10.5.

$$
\begin{array}{r}
1.5 \\
\times\,7 \\
\hline
10.5 \\
\end{array}
$$

If both numbers to be multiplied include decimals, the process is the same.

$$
\begin{array}{r}
{}^{3}\,1.5 \\
\times\,7.1 \\
\hline
15 \\
+\,1050 \\
\hline
1065 \\
\end{array}
$$

Counting the decimal places, 1.5 has one decimal place, and 7.1 has one decimal place. Therefore, the decimal in the answer is placed two spaces from the right, making the answer 10.65.

$$\begin{array}{r} \overset{3}{1.5} \\ \times\, 7.1 \\ \hline 15 \\ +\, 105\mathbf{0} \\ \hline 10.65 \end{array}$$

When zeros appear to the left of the decimal point with no other numbers, they are ignored. Zeros are often placed in the ones place to make sure the decimal point is not overlooked. For example, 0.01 is the same as .01, as the zero to the *left* of the decimal is simply a placeholder and has no value. However, zeros to the *right* of the decimal are very important, as long as another number is further to the right of those zeros. These zeros indicate the value of the number.

To multiply 0.01 times 0.006, the calculation would be as follows:

$$\begin{array}{r} 0.006 \\ \times\, 0.01 \\ \hline 0006 \\ +\, 0000\mathbf{0} \\ \hline 0.00006 \end{array}$$

The factors being multiplied have a combined total of five decimal places between them, so the product also has five decimal places. The digits to the right of the decimal in each number determine the number of decimal places in the product.

3.2.4 Division of Decimals

Dividing decimals is done exactly the same as the division of whole numbers, but the decimal point must be handled properly in the divisor, dividend, and quotient. For example, to divide 93 by 3.1, the problem would be set up as shown.

$$3.1\overline{)93}$$

Any decimal point in the divisor needs to be moved to the right to make a whole number. Therefore, the decimal point used in the divisor 3.1 needs to be moved to the far-right position, thus making the divisor 31. Since the decimal point was moved one place to the right in the divisor, the decimal point in the dividend needs to be moved the same number of times in the same direction. In this way, the dividend 93 becomes 930. The zero had to be added to the 93 as a placeholder.

$$3.1.\overline{)93.0.}$$

The division can now be completed as whole numbers. The decimal point in the quotient will be directly above the decimal point in the dividend.

$$\begin{array}{r} 30. \\ 31.\overline{)930.} \\ -\,930 \\ \hline 0 \end{array}$$

$$93 \div 3.1 = 30$$

While division using whole numbers can result in answers with remainders beside the quotient, division using decimals can express remainders as a decimal portion of the quotient. First, let us review whole number division by dividing 8 by 3.

$$
\begin{array}{r}
2 \\
3\overline{)8} \\
-6 \\
\hline
2
\end{array}
$$

$$
= 2\ R2
$$

The whole number 3 does not divide evenly into 8, so there is a remainder of 2. This problem's answer can be expressed differently using decimals. A decimal point is added after the 8, and a 0 is written to the right of the decimal point. The 0 is then carried down to the remaining 2, making it 20. A decimal point is also placed in the quotient line in the same place position. Thus, it is just to the right of the 2.

$$
\begin{array}{r}
2. \\
3\overline{)8.0} \\
-6 \downarrow \\
\hline
2\,0
\end{array}
$$

Now 3 can be divided into 20, with 6 as the result. The 6 is placed to the right of the decimal place above the 0. The 6 is now multiplied by the 3, making 18. The 18 is written below the 20 and subtracted from 20, leaving 2.

$$
\begin{array}{r}
2.6 \\
3\overline{)8.0} \\
-6 \downarrow \\
\hline
2\,0 \\
-1\,8 \\
\hline
2
\end{array}
$$

Another 0 is then added to the right of the decimal point and carried down to the remaining 2 once again. Now 3 can be divided into 20, with 6 as the result and a remainder of 2.

$$
\begin{array}{r}
2.66 \\
3\overline{)8.00} \\
-6 \downarrow\ \downarrow \\
\hline
2\,0 \\
-1\,8 \downarrow \\
\hline
2\,0 \\
-1\,8 \\
\hline
2
\end{array}
$$

If two decimal places are sufficient for the accuracy of the answer, the division process is stopped at this point. If it is desired to continue to perhaps 3 or 4 decimal places, continue following the same procedure. However, as seen in this example, continuing beyond two decimal places will serve no purpose, as the 6 will repeat endlessly. Thus, the quotient can be stated as $2.\overline{66}$. The line above the sixes indicates that it repeats. This answer could also be rounded up to 2.67.

Review Questions

Name _____ **Date** _____ **Class** _____

Fraction to Decimal Conversion Exercises

Exercise 3-1

Calculate the decimal equivalent of the following fractions. Round answers to the third decimal place (thousandths).

1. $\frac{3}{10}$ = _____

2. $\frac{5}{1,000}$ = _____

3. $\frac{7}{9}$ = _____

4. $\frac{8}{12}$ = _____

5. $\frac{5}{6}$ = _____

6. $\frac{11}{12}$ = _____

7. $\frac{9}{11}$ = _____

8. $\frac{6}{7}$ = _____

9. $\frac{21}{22}$ = _____

10. $\frac{47}{50}$ = _____

11. $\frac{12}{13}$ = _____

12. $\frac{1}{24}$ = _____

Exercise 3-2

Using the decimal equivalents as a guide, arrange the fractions in Exercise 3-1 in order from smallest value to largest value.

1. _____
2. _____
3. _____
4. _____
5. _____
6. _____
7. _____
8. _____
9. _____
10. _____
11. _____
12. _____

Decimal to Fraction Conversion Exercises

Exercise 3-3

Convert the following decimals to fractions, and reduce to the lowest terms.

1. 0.34 = _____

2. 0.275 = _____

3. 0.7 = _____

4. 0.0356 = _____

5. 0.074 = _____

6. 0.102 = _____

7. 0.0181 = _____

8. 0.23 = _____

9. 0.190 = _____

10. 0.668 = _____

11. 0.0034 = _____

12. 0.0002 = _____

Practical Exercise 3-4

The monthly energy usage in kilowatt-hours (kWh) for several appliances is listed below.

1. Calculate the monthly cost of operating each of the appliances. The electric utility charges $0.142 per kilowatt-hour. Round answers to the nearest hundredth.

    A. Television = 10.8 kWh _____

    B. Freezer = 90 kWh _____

    C. Refrigerator = 50 kWh _____

    D. Range = 65 kWh _____

    E. Lights = 22 kWh _____

    F. Electric Water Heater = 420 kWh _____

    G. Garage Door Opener = 5.6 kWh _____

    H. Dishwasher = 25 kWh _____

Name _____ **Date** _____ **Class** _____

    I.   Washing Machine = 26.5 kWh _____

    J.   Clothes Dryer = 45 kWh _____

    K. Computer = 6 kWh _____

    L.   Vacuum = 2.2 kWh _____

    M. Coffee Maker = 7.2 kWh _____

    N. Microwave Oven = 7.6 kWh _____

2. Add the totals together to determine the total energy bill for the month.

_____

Decimal Exercises

Practical Exercise 3-5

A school is planning to retrofit classroom lighting from fluorescent to LED. This is to improve the amount of light, eliminate maintenance costs of ballasts and lamps, and reduce energy costs. The existing fluorescent luminaires draw 0.96 amperes each. The new LED luminaires will draw 0.46 amperes each. There are 24 classrooms, each having 16 fluorescent luminaires. Round answers to the nearest hundredth.

1. What is the amount of current drawn by the existing luminaires in one of the classrooms?

   _____ amperes

2. What is the total amount of current drawn by the existing luminaires in all of the classrooms?

   _____ amperes

3. What is the amount of current that will be drawn by the new LED luminaires in one of the classrooms?

   _____ amperes

4. What is the total amount of current that will be drawn by the new LED luminaires in all of the classrooms?

   _____ amperes

5. How much will the new LED luminaires reduce the current drawn by lighting in one of the classrooms?

   _____ amperes

6. How much will the new LED luminaires reduce the total current drawn by lighting in all of the classrooms?

   _____ amperes

Practical Exercise 3-6

Section 210.20(A) of the *National Electrical Code* limits the amount of current on an overcurrent device to 80% (0.8) of its maximum current rating if it feeds a continuous load. Lighting in a school is an example of a continuous load, as it draws the maximum current for three hours or more. The lighting branch circuits in a school will be fed with 12 American Wire Gauge (AWG) branch-circuit conductors that are protected by 20-ampere circuit breakers.

1. What is the maximum amount of current permitted on one of the lighting branch circuits?

   _____ amperes

Practical Exercise 3-7

An electrician has been sent to a restaurant to wire a new walk-in freezer. The food from the old freezer has been put into coolers while the freezer is being replaced. In an effort to prevent the food from defrosting, the electrician and the refrigeration technician are working together to get the freezer replaced as quickly as possible. Once the new unit is connected and running, the temperature inside the freezer begins to drop at a rate of 1.3°F (0.73°C) over a 10-minute period.

1. If this rate of decrease continues, how much will the box temperature have dropped in one hour?

   _____

2. To protect the product quality, the box temperature needs to be maintained at –20°F (–28.9°C). When the new freezer was started, the temperature inside the freezer was 70°F (21.1°C). Assuming the same rate of temperature decrease, approximately how long will it take for the unit to reach the desired temperature?

   _____

Practical Exercise 3-8

A pre-fabrication electrician for an electrical contractor works in the shop every day building electrical components to be installed in the field. The time clock that records the hours worked measures time in hours and tenths of an hour. The timecard for the fabricator shows the following hours worked for the week:

- Monday: 7.2 hours
- Tuesday: 8.3 hours
- Wednesday: 9.1 hours
- Thursday: 8.9 hours
- Friday: 9.2 hours

1. How many hours did the fabricator work this week?

   _____ hours

Name _____   **Date** _____   **Class** _____

2. The rate of pay for this employee is $16.45 per hour for regular pay, with overtime (over 40 hours per week or 10 hours in one day) paid at a rate of 1 1/2 times the regular rate. What is the amount that is due to this employee for the week?

   $_____

3. This employee participates in a retirement plan that allows a portion of the pay to be placed in a separate account. The plan stipulates that 3.5% (0.035) of the employee's pay will be set aside for retirement. How much of this week's paycheck will be placed in the retirement account?

   $_____

Practical Exercise 3-9

A large building has five zones of floor heat. Each zone has a different wattage heating element appropriately sized for the area of the zone. The wattages of the heating elements are listed below.

Zone 1: 1,000 watts   Zone 4: 2,500 watts

Zone 2: 750 watts   Zone 5: 500 watts

Zone 3: 1,500 watts

1. If all the heating elements were on at the same time for one hour, how much energy (watt/hour) would be consumed?

   _____

2. What is the average value of the heating elements?

   _____

   Over the past 24 hours, it was recorded that the heating elements ran for the following amounts of time:

Zone 1: 6.8 hours   Zone 4: 14.7 hours

Zone 2: 9.3 hours   Zone 5: 18.1 hours

Zone 3: 12.5 hours

3. How many watt/hours of energy were consumed by each of the five heating elements over the past 24 hours?

   Zone 1: _____   Zone 4: _____

   Zone 2: _____   Zone 5: _____

   Zone 3: _____

4. The current cost for electricity is $0.0002 per watt/hour. What was the energy cost to operate each heating element for this 24-hour period? Round answers to the nearest hundredth.

   Zone 1: _____   Zone 4: _____

   Zone 2: _____   Zone 5: _____

   Zone 3: _____

(Continued)

5. What is the total energy cost for the 24-hour period?

_____

Practical Exercise 3-10

In an effort to catch up on backed-up service calls, the owner asked the service electricians to work a long day before leaving for an extended holiday weekend. One of the electricians worked 13 hours that day and completed five service calls.

1. If each service call lasted the same amount of time, how long did one service call last? Include the travel time between calls with each service call. Calculate to the nearest 1/10 of an hour (one decimal place).

_____

  Responding to the grateful customers and the good work of the electricians, the owner of the service company decided to pay everyone at 1.67 times their hourly rate for that long day.

2. If an apprentice normally earns $19.50 an hour and worked a 10-hour shift at 1.67 times the normal rate that day, how much would the apprentice take home for one day's work?

_____

Practical Exercise 3-11

A service electrician leaves the shop at 7:00 a.m. to go to the first call of the day. The travel time is 30 minutes. After completing the service work, travel to the next job takes 40 minutes. After that call, the electrician takes a 30-minute lunch break. Travel to the final call of the day takes 20 minutes. Once the day's work is done, the electrician drives for 30 minutes to get back to the shop at 3:00 p.m.

1. How much total time (in minutes) did the electrician spend working (not traveling or eating lunch)?

_____ minutes

2. How much total time (in hours) did the electrician spend working on service calls (not traveling or eating lunch)?

_____ hours

3. If each service call lasted the same amount of time, how long (in minutes) would a service call last?

_____ minutes

4. If each service call lasted the same amount of time, how long (in hours) would a service call last? Calculate and round to the nearest 1/100 of an hour (two decimal places).

_____ hours

CHAPTER 4

Percentages, Ratios, and Proportions

Objectives

Information in this chapter will enable you to:

- Convert percentages to decimals.
- Convert decimals to percentages.
- Calculate addition, subtraction, multiplication, and division of percentages.
- Explain how ratios are used in the electrical industry.
- Explain how proportions are used in the electrical industry.

Technical Terms

extremes	percentage	ratio
means	proportion	

4.1 Percentages, Ratios, and Proportions

When comparing values of data, understanding the use of percentages, ratios, and proportions is extremely helpful to the tradesperson. Percentages are used in calculating raceway fill; conductor ampacities; continuous loads; power factor; and the profit, loss, and tax rates on job pricing. Ratios and proportions are used in electrical theory calculations, transformer calculations, and on the jobsite when mixing concrete or sealing compounds.

A **percentage** is a portion of a whole, just like a fraction. A percentage is expressed by using the percent sign (%) to indicate the value of that portion. Because a whole is represented as 100%, any portion thereof will be a lesser value.

For example, 50% represents a portion equal to 1/2 of the whole, as 50 is exactly one-half of one hundred. Thus, if a conductor is carrying 50% of its maximum current, it is carrying exactly one-half of the amount of current it is capable of.

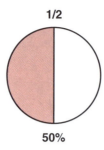

Percentages can also be expressed as *greater than the whole*. A value representing twice as much as the original whole number could be said to equal 200% or two times 100%.

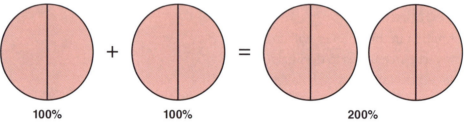

A **ratio** is an expression of a comparison of two or more values. Ratios are shown as two numbers separated by a colon (:). For example, if a journeyman electrician can install receptacles four times faster than the new apprentice, that means the journeyman has installed four receptacles every time the apprentice has installed one. The journeyman to apprentice production ratio is 4:1.

Journeyman/Apprentice

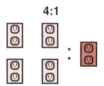

A **proportion** is a comparison of two ratios. In the proportion, the ratios are separated by an equals sign (=). An example of proportion is a fraction with high values and the same fraction reduced to its lowest terms.

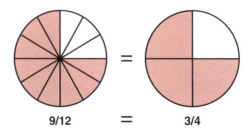

Even though these two fractions have different numerators and denominators, they are equal because they maintain the same numerator to denominator ratio. They are proportional to each other. Proportions are often used as a means to determine an unknown value.

$$\frac{7}{60} = \frac{49}{?}$$

4.1.1 Converting Percentages to Decimals

In order to perform mathematical calculations involving percentages, it is best to first convert the percentage to decimal format. Remember, a percentage is only a portion of a whole (100% or 1.0). Therefore, a percentage can be written as a fraction and then simply converted to decimal form. When writing a percentage as a fraction, the denominator will always be 100. For example, 50% can be written as 50/100, reduced to 5/10, and then converted to the decimal format as 0.5.

$$50\% = \frac{50}{100}$$

$$\frac{50}{100} = \frac{5}{10}$$

$$\frac{5}{10} = 0.5$$

Likewise, 25% can be written as 25/100, and since it cannot be reduced by 10, it can simply be converted to 0.25. This example shows that the easiest way to turn a percentage into a decimal is to turn the percent sign into a decimal and move it two places to the left.

$$25\% = 0.25\%$$

Values greater than 100% are handled in the same manner, but the decimal format will be a number greater than 1, rather than a number less than 1. For example, consider 150%.

$$150\% = 1.50\%$$

Steps for Converting a Percentage to a Decimal
1. Delete the percent sign.
2. Move the decimal point two places to the left.

Percentage to Decimal Conversion

Percentage	Symbol Swap and Shift	Decimal
87%	87.	0.87

Goodheart-Willcox Publisher

Once a percentage has been converted to decimal format, calculations can be handled as decimals.

4.1.2 Using Percentages

In the electrical industry, percentages are often used when performing calculations such as ampacity, raceway fill, and power factor, as well as when doing paperwork such as estimating, billing, etc. The use of the percent sign gives an immediate reference point, as the value expressed is always shown in relation to 100%.

For example, if the voltage drop on the conductors feeding a motor is 3%, it can be readily determined that the motor is receiving the remaining portion (97%) of the applied voltage.

$$100\% - 3\% = 97\%$$

In more complex calculations, it is easier to convert the percentage to a decimal and then do the arithmetic by following the rules of using decimals. Thus, values will be expressed in both the percentage form and the decimal form, depending on the situation and the calculations necessary.

The *National Electrical Code* has several requirements that involve multiplying by percentages. A common example is when a branch circuit feeds a continuous load. A continuous load is a load that is expected to be on for three hours or more. Being on for a sustained amount of time will lead to heating up of the conductors, terminals, overcurrent device, and so forth. *Section 210.20(A)* of the *NEC* requires that conductors and overcurrent devices that feed a continuous load have an ampacity of not less than 125% of the continuous load. To determine the minimum ampacity of a conductor with a continuous load, multiply the current value by the decimal equivalent of the percentage, which is 1.25 (125%).

For example, when trying to find the minimum ampacity of branch-circuit conductors that feed a 42-ampere continuous load, you would multiply the current value (42) by the decimal equivalent of the percentage (1.25).

$$42 \text{ amperes} \times 1.25 \ (125\%) = 52.5 \text{ amperes}$$

You will also need to perform continuous load calculations from the opposite perspective, such as when determining the maximum amount of continuous load current that is permitted on a specific conductor or overcurrent device. In this case, you will use division rather than multiplication. To determine the maximum amount of continuous load a breaker can carry, divide the ampacity of the breaker by the decimal equivalent of the percentage, which is 1.25 (125%).

For example, when trying to find the maximum amount of continuous load a 30-ampere breaker can carry, you would divide the ampacity (30) by the decimal equivalent of the percentage (1.25).

$$30 \text{ amperes} \div 1.25 \ (125\%) = 24 \text{ amperes}$$

4.2 Using Ratios and Proportions

A ratio is an expression of a comparison of two or more values. Ratios are shown as two numbers separated by a colon (:). A proportion is a comparison of two or more ratios. A proportion is most often shown as two or more ratios separated by an equals sign (=).

4.2.1 Comparing Numbers Using Ratios

Ratios are often used to compare the values of components in a mixture. A good example is when mixing chemical cleaners. Mixing chemicals isn't something that an electrician will be doing every day, but it is a great way to illustrate ratios and proportions.

The chemical cleaner label might instruct the user to make a 15:1 solution using tap water. This means that for every ounce of chemical cleaner used, 15 ounces of water would be added.

4.2.2 Comparing Ratios Using Proportions

Proportions are a means of comparing ratios. Ratios can be used to find unknown values of equal proportions. For example, if the cleaning solution mentioned earlier was mixed at a 15:1 concentration, each ounce of chemical would be mixed with 15 ounces of water. This would make a diluted solution of 16 ounces. A proportion can be used to find the right amount of chemical to use to make one gallon of mix.

Ratio of water to chemical: 15:1

Desired amount of solution: 1 gallon (128 ounces)

To solve for the unknown quantity, use a simple algebraic formula. In this proportion, use an X to represent the unknown quantity:

Total parts of solution: Parts of chemical =

Total ounces of solution: Number of ounces of chemical

To calculate the total sum of the solution, add the two ingredients (chemical part plus water parts):

15 parts water + 1 part chemical = 16 parts solution

Therefore, the proportion would be written as:

16:1 = 128:X

To solve for the unknown quantity (X), the two *means* are multiplied by each other, and the two *extremes* are multiplied by each other. Note that means and extremes can be seen in proportions expressed as ratios or fractions. The **means** are the two numbers closest to the equals sign in a ratio. The **extremes** are the two numbers farthest from the equals sign in a ratio. Means and extremes can also be expressed in proportional fractions.

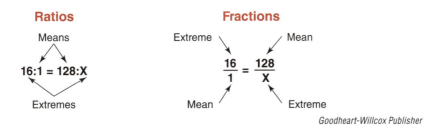

Goodheart-Willcox Publisher

In order to solve for the unknown, the first step is to multiply. Start by multiplying the means together.

$1 \times 128 = 128$

Now multiply the extremes.

$16 \times X = 16X$

This leaves us with an unknown quantity adjacent to a whole number. This means 16 multiplied by the unknown value would equal 128.

$16X = 128$

To solve for the unknown, the X must be isolated. To do that, both sides of the equation are divided by 16.

$$\frac{\cancel{16}X}{\cancel{16}} = \frac{128}{16}$$

Much like reducing a fraction to its lowest terms, the 16 above and the 16 below cancel out each other, leaving only the X. Dividing 128 by 16 will reveal the value of the unknown.

$$X = \frac{128}{16} = 8$$

X is usually written simply as X, as the multiplier 1 is assumed. Thus X = 8.

For a 15:1 solution in the amount of 1 gallon (128 ounces), you would add 8 ounces of chemical to 120 ounces of water to make one gallon (128 ounces).

$$15:1 = 120:8$$

These ratios can be checked. Arrange each ratio as a fraction.

$$\frac{15}{1} = \frac{120}{8}$$

Divide the higher value numerator by the other numerator.

$$120 \div 15 = 8$$

Divide the higher value denominator by the other denominator.

$$8 \div 1 = 8$$

The quotients are the same because the two ratios are equal. Proportions can be viewed as two forms of the same fraction. One of these fractions is in a more reduced form (15/1) than the other fraction (120/8).

$$\frac{15}{1} = \frac{120}{8}$$

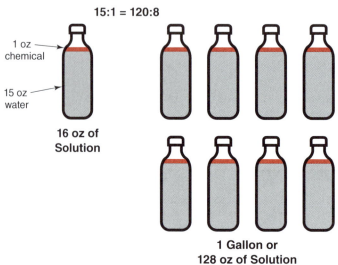

Proportions

15:1 = 120:8

1 oz chemical

15 oz water

16 oz of Solution

1 Gallon or 128 oz of Solution

Goodheart-Willcox Publisher

4.2.3 Transformer Ratios and Proportions

A *transformer* is an electrical device that contains two sets of windings, primary and secondary, that are electrically isolated from one another but are physically wrapped around a common magnetic core.

The primary winding is where the applied voltage connects to the transformer, and the secondary winding is where the transformed voltage is available to connect to a load. The secondary voltage can be stepped up from the voltage applied to the primary, or it can be stepped down. When ac voltage is applied to the primary side of a transformer, magnetic induction causes a voltage to be induced into the secondary coil of the wire. If the secondary coil is connected to a completed circuit, current will flow.

The applied voltage has a direct relationship to the secondary voltage based on the number of turns in the primary compared to the secondary. A transformer that has 20 windings on the primary and 5 windings on the secondary has a 20:5, which can be reduced to 4:1. The voltage applied to the primary of a transformer has a direct relationship to the number of windings in the secondary. The following formula can be used to set up the proportion when solving for voltage.

$$\frac{E_P}{E_S} = \frac{N_P}{N_S}$$

where

E_P = primary voltage

E_S = secondary voltage

N_P = number of turns in primary

N_S = number of turns in secondary

The transformer pictured has a 4:1 turns ratio with an applied voltage of 120 volts. To solve the ratio, enter the known values and cross multiply.

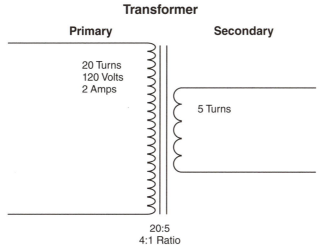

Transformer

Primary   Secondary

20 Turns
120 Volts
2 Amps

5 Turns

20:5
4:1 Ratio

Goodheart-Willcox Publisher

$$\frac{120}{E_S} \diagup\!\!\!\!\times \frac{4}{1}$$

$$4E_S = 120$$

$$E_S = 30 \text{ volts}$$

A transformer's current values have an inversely proportional relationship to its turns ratio and voltage ratio. If the secondary voltage of a transformer has been stepped up from the primary, then the current on the secondary will be less than the primary current. If the secondary voltage has been stepped down from the primary, then the current on the secondary will be higher than the primary current. The following formulas can be used to set up the proportion when solving for current.

$$\frac{N_P}{N_S} = \frac{I_S}{I_P}$$

where

N_P = number of turns in primary

N_S = number of turns in secondary

I_S = current in secondary

I_P = current in primary

or

$$\frac{E_P}{E_S} = \frac{I_S}{I_P}$$

E_P = primary voltage

E_S = secondary voltage

I_S = current in secondary

I_P = current in primary

The transformer pictured earlier has a 4:1 turns ratio with a primary current of 2 amperes. To solve the ratio, enter the known values and cross multiply.

$$\frac{N_P}{N_S} = \frac{I_S}{I_P}$$

$$\frac{4}{1} \diagtimes \frac{I_S}{2}$$

$$I_S = 8 \text{ amperes}$$

or

$$\frac{E_P}{E_S} = \frac{I_S}{I_P}$$

$$\frac{120}{30} \diagtimes \frac{I_S}{2}$$

$$I_S = 8 \text{ amperes}$$

Review Questions

Conversion Exercise

Exercise 4-1

Convert the following percentages to decimal format.

1. 13% = _____
2. 37% = _____
3. 93% = _____
4. 10% = _____
5. 100% = _____
6. 136% = _____

7. 9% = _____
8. 419% = _____
9. 1.2% = _____
10. 75.3% = _____
11. 110.1% = _____
12. 57.3% = _____

Percentage Exercises

Exercise 4-2

Complete the following percentage calculations.

1. 90% – 12% = _____
2. 103% – 71% = _____
3. 21% – 11% = _____
4. 156% – 142% = _____
5. 57% + 43% = _____
6. 26% + 87% = _____

7. 89% × 10% = _____
8. 152% × 50% = _____
9. 32% × 1% = _____
10. 90% ÷ 3 = _____
11. 90% ÷ 3% = _____
12. 32% ÷ 2 = _____

Practical Exercise 4-3

1. An industrial facility has a power factor of 73%. After installing power factor correction equipment, the power factor improved to 96%. By what percentage did the power factor improve?

_____

2. The *National Electrical Code* requires branch-circuit conductors that feed a continuous load to have a minimum ampacity of 125% of the full-load current. What is the minimum ampacity of a conductor that feeds a 16-ampere continuous load?

_____

3. The *National Electrical Code* permits an inverse time breaker, that is used for short circuit and ground fault protection, to be rated at 250% of the full-load current of a single-phase motor. What is the maximum size inverse time breaker permitted for a motor with a full-load current of 80 amperes?

_____

4. The *National Electrical Code* requires a branch-circuit overcurrent protection device to be rated at 125% of a continuous load. If the branch-circuit overcurrent device is rated at 15 amperes, what is the maximum amount of continuous load current permitted on the circuit?

_____

5. The *National Electrical Code* requires conductors that feed a capacitor to be not less than 135% of the rated current of the capacitor. What is the minimum ampacity of the conductors if the capacitor has a rated current of 56 amperes?

_____

Ratio and Proportion Exercises

Exercise 4-4

Solve the following proportions for the unknown factor.

1. 10:1 = 100:X

_____

2. 7:3 = 21:X

_____

3. 25:1 = 50:X

_____

4. 5:2 = X:4

_____

5. 4:7 = X:28

_____

6. 3:11 = X:77

_____

7. X:7 = 4:1

_____

8. X:1 = 63:7

_____

9. X:4 = 9:6

_____

10. 3:X = 18:12

_____

11. 14:X = 7:6

_____

12. 1:X = 41:41

_____

Practical Exercise 4-5

A single-phase step-down control transformer has a primary-to-secondary winding ratio of 10:1.

1. If the primary winding voltage is 240 volts, what is the secondary winding voltage?

_____

2. If the secondary winding voltage is 12 volts, what is the primary winding voltage?

_____

3. If the secondary current is 2 amperes, what is the primary current?

_____

4. If the primary current is 3 amperes, what is the secondary current?

_____

CHAPTER 5

Linear Measurements and Conversions

Objectives

Information in this chapter will enable you to:

- Convert US Customary linear measurements between feet and inches.
- Calculate addition, subtraction, multiplication, and division of US Customary linear measurements.
- Recognize the non-electrical measurements used by the construction and maintenance industries.
- Convert between different units in the US Customary and SI systems.

Technical Terms

air velocity
air volume
British thermal unit (Btu)
Btu per hour (Btu/hr)
calorie (cal)
capacity
Celsius scale
cubic feet per minute (cfm)
cubic meter per minute (m³/min)
degree

Fahrenheit scale
feet per minute (fpm)
foot (ft)
gram (g)
heat
inch (in)
inches of water column (in. WC)
joule (J)
Kelvin scale
kilogram (kg)
kilojoule (kJ)
kilometer (km)

kilopascal (kPa)
linear measurement
mass
meter (m)
meter per minute (m/min)
micron (μm)
millimeter (mm)
ounce (oz)
pascal (Pa)
pound (lb)
pounds per square inch (psi)

pressure
Rankine scale
SI system
square inch (in²)
square meter (m²)
temperature
therm
ton (t)
ton of refrigeration
US Customary system
weight
yard (yd)

5.1 Introduction to Measurement

This chapter will cover non-electrical measurements and conversions. Electrical measurements will be covered in later chapters. The proper use and understanding of measurements is necessary for satisfactory performance in the electrical and mechanical trades. Errors in measurement can lead to wasted material and miscalculations, resulting in lost time and money for the service technician and the employer or building owner.

Some of the measurements covered in this chapter, such as linear measurement, are used nearly every day by electrical workers, while others are rarely encountered. Although not common to every electrician, specialized portions of the electrical industry, such as instrumentation, will work with many of the measurements covered. Additionally, electricians are making electrical connections to equipment and work alongside other trades that regularly use these measurements. It is helpful to be familiar with their systems and understand the workings of the equipment in order to choose the correct wiring methods and to be able to communicate effectively.

In the United States, two systems of linear measurement are encountered on a daily basis. The **US Customary system**, also referred to as the *inch-pound (IP) system*, consists of centuries-old measurement units, such as the inch, gallon, and pound.

The *International System of Units (SI)* is commonly called the **SI system** or *metric system*. It is based on standard units for length, weight, and volume with prefixes in powers of ten. This system was designed to replace multiple units of measurement with values that could be used worldwide.

5.2 Linear Measurement

Goodheart-Willcox Publisher

Linear measurement is used to determine the length, width, or depth of an object or the distance between two points. Linear units are commonly used while installing electrical systems.

5.2.1 US Customary

In the US Customary system, linear measurement is based on the **inch (in)**, which can be broken into fractions for precise measurement accuracy.

Fractions of an Inch

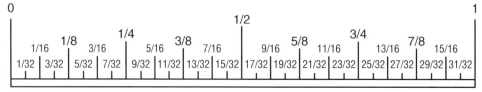

Goodheart-Willcox Publisher

In this system, a length of 12 inches is called a **foot (ft)**, and a length of 3 feet is known as a **yard (yd)**. Symbols are often used to denote inches and feet. The symbol for inches is ", while ' is the symbol for feet. A length of 6 feet and 8 inches is written as 6'-8".

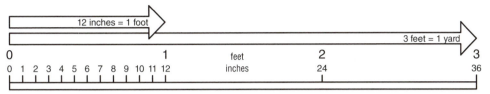

Linear Measurement–US Customary Units

Goodheart-Willcox Publisher

Typically, field measurements use feet, inches, and fractions of inches for laying out locations and determining lengths. Yards are seldom used for linear measurement in construction.

5.2.2 Converting Linear Measurements

When working with linear measurements, electricians often convert measurements back and forth between feet and inches. This is a useful skill that can make addition, subtraction, multiplication, and division of linear measurements easier.

Converting Feet to Inches

Converting a measurement from feet to inches is accomplished by multiplying the number of feet by 12. For example, to convert 4 feet to inches:

$$4' \times 12 = 48''$$

If the measurement to be converted contains both feet and inches, it becomes a two-step process.
1. Multiply the number of feet by 12.
2. Add the number of inches to the product of Step 1.

For example, to convert 6'-8" to inches:

$$6' \times 12 = 72''$$

Then, add 8" to 72".

$$72'' + 8'' = 80''$$

Converting Inches to Feet

Converting a measurement from inches to feet is accomplished by dividing the number of inches by 12. For example, to convert 84" to feet:

$$
\begin{array}{r}
7' \\
12\overline{)84''} \\
-84 \\
\hline
0
\end{array}
$$

The number of inches to be converted to feet may not be evenly divisible by 12. When this is the case, you will end up with a remainder. The remainder is kept as inches to complete the conversion process. For example, to convert 106 inches to feet:

$$
\begin{array}{r}
8' \\
12\overline{\smash{\big)}106''} \\
-96 \\
\hline
10''
\end{array}
$$

Then, keeping the remainder 10 in inches, the answer would be 8'-10".

5.2.3 Adding and Subtracting Feet and Inches

Adding and subtracting measurements that contain feet and inches is a common task when working on construction sites. There are two methods to add and subtract measurements. The first method involves converting the measurements into inches and then performing addition or subtraction. The second method involves directly adding the number of feet together followed by adding the number of inches together.

Adding Feet and Inches (Method 1)

There are three steps for adding feet and inches using Method 1. Step 3 will not be necessary if leaving the answer in inches is sufficient.

1. Convert the feet and inches measurement into inches.
2. Add the number of inches from Step 1 together.
3. (Optional) Convert the sum from Step 2 back to feet and inches.

For example, to add 3'-6" to 6'-2", first convert to inches.

$$3'\text{-}6''$$
$$3' \times 12 = 36''$$
$$36'' + 6'' = 42''$$

$$6'\text{-}2''$$
$$6' \times 12 = 72''$$
$$72'' + 2'' = 74''$$

Then add the two conversions together.

$$42'' + 74'' = 116''$$

Finally, if necessary, convert 116" back to feet.

$$
\begin{array}{r}
9' \\
12\overline{\smash{\big)}116''} \\
-108 \\
\hline
8''
\end{array}
$$

This results in an answer of 9'-8".

Adding Feet and Inches (Method 2)

There are four steps for adding feet and inches using Method 2. If the result from the number of inches is less than 12, then only Steps 1 and 2 are necessary.

1. Add the number of feet together.
2. Add the number of inches together.
3. (Optional) Convert the sum from Step 2 into feet and inches.
4. (Optional) Add the sum from Step 1 to the number of feet and inches from Step 3.

For example, to add 2'-5" to 3'-3", first add the number of feet.

$$\begin{array}{r} 2'\text{-}5'' \\ +\ 3'\text{-}3'' \\ \hline 5' \end{array}$$

Next, add the number of inches.

$$\begin{array}{r} 2'\text{-}5'' \\ +\ 3'\text{-}3'' \\ \hline 5'\text{-}8'' \end{array}$$

Since the number of inches is less than 12, you can skip Steps 3 and 4.
To add 4'-8" to 12'-5", first add the number of feet.

$$\begin{array}{r} 4'\text{-}8'' \\ +\ 12'\text{-}5'' \\ \hline 16' \end{array}$$

Next, add the number of inches.

$$\begin{array}{r} 4'\text{-}8'' \\ +\ 12'\text{-}5'' \\ \hline 16'\text{-}13'' \end{array}$$

Since the number of inches is more than 12, convert the inches to feet.

$$\begin{array}{r} 1' \\ 12\overline{)13''} \\ -\ 12 \\ \hline 1'' \end{array}$$

Add 1'-1" to 16'.

$$\begin{array}{r} 1'\text{-}1'' \\ +\ 16' \\ \hline 17'\text{-}1'' \end{array}$$

Subtracting Feet and Inches (Method 1)

There are three steps to subtracting feet and inches using Method 1. Step 3 will not be necessary if leaving the answer in inches is sufficient.

1. Convert the feet and inches measurements into inches.
2. Subtract the smaller number of inches from the larger number.
3. (Optional) Convert the difference from Step 2 into feet and inches.

For example, to subtract 4'-6" from 12'-2", first convert to inches.

$$4'\text{-}6''$$
$$4' \times 12 = 48''$$
$$48'' + 6'' = 54''$$

$$12'\text{-}2''$$
$$12' \times 12 = 144''$$
$$144'' + 2'' = 146''$$

Then subtract the smaller number from the larger.

$$
\begin{array}{r}
146'' \\
- \ 54'' \\
\hline
92''
\end{array}
$$

Finally, convert back to feet and inches.

$$
\begin{array}{r}
7' \\
12\overline{)92''} \\
- \ 84 \\
\hline
8''
\end{array}
$$

The converted answer is 7'-8".

Subtracting Feet and Inches (Method 2)

There are three steps to subtracting feet and inches using Method 2. Step 1 will not be necessary if the top number of inches is larger than the bottom number.

1. Subtract 1 from the number of feet and add 12 to the number of inches (if necessary).
2. Subtract the number of inches.
3. Subtract the number of feet.

For example, to subtract 5'-6" from 10'-8", first subtract the number of inches.

$$
\begin{array}{r}
10'\text{-}8'' \\
- \ 5'\text{-}6'' \\
\hline
2''
\end{array}
$$

Then subtract the number of feet.

$$
\begin{array}{r}
10'\text{-}8'' \\
- \ 5'\text{-}6'' \\
\hline
5'\text{-}2''
\end{array}
$$

To subtract 8'-10" from 20'-4", start by subtracting 1 from 20' and adding 12 to 4".

$$
\begin{array}{rr}
20' & 4'' \\
- \ 1 & + \ 12 \\
\hline
19' & 16''
\end{array}
$$

Subtract the number of inches.

$$
\begin{array}{r}
19'\text{-}16'' \\
- \ 8'\text{-}10'' \\
\hline
6''
\end{array}
$$

Finally, subtract the number of feet.

$$
\begin{array}{r}
19'\text{-}16'' \\
-\ 8'\text{-}10'' \\
\hline
11'\text{-}6''
\end{array}
$$

5.2.4 Multiplying and Dividing Feet and Inches

When multiplying and dividing feet and inches, it is best to start by converting the measurements into inches. The answer can be converted back into feet and inches after, if necessary.

Multiplying Feet and Inches

1. Convert into inches.
2. Multiply.
3. (Optional) Convert the product from Step 2 into feet and inches.

For example, to multiply 7'-3" by 5, first convert to inches.

$$7' \times 12 = 84''$$

$$84'' + 3'' = 87''$$

Next, multiply 87" by 5.

$$87'' \times 5 = 435''$$

Now, if necessary, convert back to feet and inches.

$$
\begin{array}{r}
36' \\
12\overline{)435''} \\
-\ 432 \\
\hline
3''
\end{array}
$$

This results in an answer of 36'-3".

Dividing Feet and Inches

1. Convert into inches.
2. Divide.
3. (Optional) Convert the quotient from Step 2 into feet and inches.

For example, to divide 12'-8" by 8, first convert to inches.

$$12' \times 12 = 144''$$

$$144'' + 8'' = 152''$$

Next, divide 152" by 8.

$$
\begin{array}{r}
19'' \\
8\overline{)152''} \\
-\ 8 \\
\hline
72 \\
-\ 72 \\
\hline
0
\end{array}
$$

If necessary, convert back to feet and inches.

$$\begin{array}{r} 1' \\ 12\overline{\smash{\big)}19''} \\ -12 \\ \hline 7'' \end{array}$$

This results in an answer of 1'-7".

5.2.5 SI

SI linear measurement is based on the **meter (m)**, which is spelled *metre* in all English-speaking countries except the United States. The method for determining the length of the meter has changed over the past few centuries, but now it has been standardized and accepted throughout the world.

The SI, or metric, system works on the decimal system. Thus, powers of ten are used to identify lengths that are shorter and longer than the meter. For example, 1/1,000 of a meter is called a **millimeter (mm)**, while 1,000 meters is called a **kilometer (km)**.

Linear Measurement–SI Units

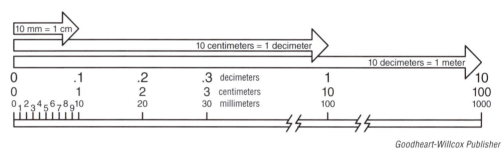

Goodheart-Willcox Publisher

Prefixes used in SI measurement are the same throughout the system, whether used for linear measurement, weight, or volume.

Common SI Multiples for Meter (m)

Multiple	1/1,000 meter	1/100 meter	1/10 meter	1 meter	10 meters	100 meters	1,000 meters
Name	millimeter	centimeter	decimeter	meter	dekameter	hectometer	kilometer
Symbol	mm	cm	dm	m	dam	hm	km

Goodheart-Willcox Publisher

5.3 Capacity Measurement

The amount of material a container can hold is called its **capacity**. The term *volume* is sometimes used as well and may be interchanged with the term *capacity* in common practice. However, volume is also used to describe the amount of space a three-dimensional object occupies. That term will be covered in another chapter of this text.

5.3.1 US Customary

Capacity measurement in the US Customary system is based on the fluid ounce (fl. oz.). Cups, pints, quarts, and gallons are multiples of ounces, as shown.

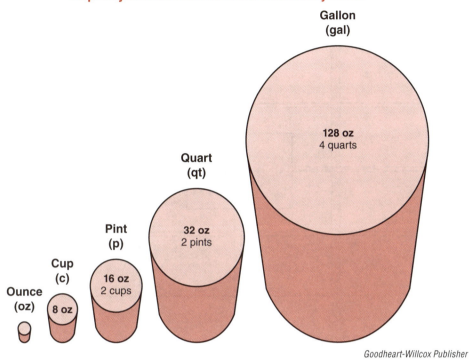

Capacity Measurements in US Customary Units

Gallon (gal)

128 oz 4 quarts

Quart (qt)

32 oz 2 pints

Pint (p)

16 oz 2 cups

Cup (c)

8 oz

Ounce (oz)

Goodheart-Willcox Publisher

5.3.2 SI

SI capacity measurement is based on the liter (l). The milliliter (ml) is 1/1,000 of a liter, and the kiloliter is 1,000 liters. Like the linear measurements in SI, the capacity measurements work on the decimal system and use prefixes added to *liter*.

5.4 Weight Measurement

Weight and mass are commonly confused terms. Though they are not the same, the terms are often used interchangeably. **Mass** may be defined as the amount of something's matter, while **weight** expresses the value of the gravitational force exerted on the matter. In other words, mass is constant, but weight depends on the force of gravity on a mass. This text will use the term *weight* rather than *mass*.

5.4.1 US Customary

In the US Customary system, weight is measured in **pounds (lb)** and **ounces (oz)**. An ounce is 1/16 of a pound, and a **ton (t)** is defined as 2,000 pounds.

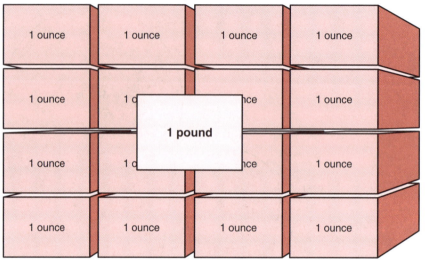

Weight Measurements in US Customary Units

16 ounces (oz) = 1 pound (lb)

2,000 pounds = 1 ton

Goodheart-Willcox Publisher

5.4.2 SI

The **kilogram (kg)** is the base unit of weight measurement in SI units. It is equal to 1,000 grams. The **gram (g)** is 1/1,000 of a kilogram.

5.5 Temperature Measurement

Temperature is the measurement of the intensity of the heat energy in a material. There are four common temperature scales: Rankine, Fahrenheit, Celsius, and Kelvin. Temperature is measured in **degrees**, which is signified by the ° symbol beside a number.

Temperature is taken into consideration when choosing wiring methods. Terminations, conductor insulation, etc. will have maximum temperature ratings. Exceeding these temperature ratings will cause damage and may lead to equipment failure.

Temperature Scales
(degrees)

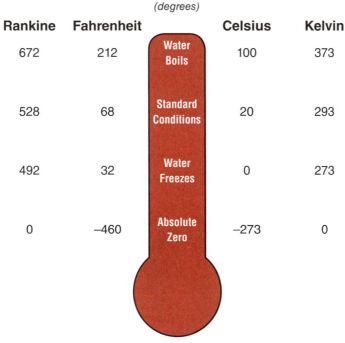

Rankine	Fahrenheit		Celsius	Kelvin
672	212	Water Boils	100	373
528	68	Standard Conditions	20	293
492	32	Water Freezes	0	273
0	−460	Absolute Zero	−273	0

Goodheart-Willcox Publisher

5.5.1 US Customary

The most common temperature scale in use in the United States is the **Fahrenheit scale**, where water at sea level and standard atmospheric pressure freezes at 32°F and boils at 212°F. The **Rankine scale** (*Fahrenheit absolute*) uses absolute zero as the baseline, and each degree has the same value as a Fahrenheit degree. The Rankine scale is used for working at ultra-low temperatures and is not typically used in the electrical industry.

5.5.2 SI

In the SI system, Celsius is used. The **Celsius scale** uses 0°C as the freezing point of water and 100°C as the boiling point. The **Kelvin scale** (*Celsius absolute*) uses absolute zero as the baseline, and each degree has the same value as a Celsius degree.

5.6 Heat Measurement

The measurement of heat energy is an important part of the HVACR industry, and it must not be confused with temperature measurement. Whereas temperature is a measure of the intensity of heat energy, **heat** is a measurement of the energy content of a material. Since electricians work closely with HVACR technicians to make electrical connections to their equipment, heat energy values will often be used to describe the heating or cooling capacity of their equipment. In the United States, the most common unit of measurement for heat is the Btu.

5.6.1 US Customary

The **British thermal unit (Btu)** is the amount of heat energy required to change the temperature of 1 lb of pure water by 1°F. When the temperature of 1 lb of pure water is raised or lowered by 1°F, the quantity of heat moved equals 1 Btu.

Since 1 Btu is a small amount of heat compared to the amounts of heat moved in some HVACR systems, a larger unit is used. The **therm** is defined as 100,000 Btu.

It is often necessary to relate the movement of the heat quantity to a time frame. Thus, the unit **Btu per hour (Btu/hr)** is commonly used. This term represents the number of Btu moved in an hour.

The term **ton of refrigeration** is also used as a measurement of heat transfer. It is based on the melting of one ton (2,000 lb) of ice at 32°F in 24 hours. Since it requires 288,000 Btu of heat to melt one ton of ice, one ton of refrigeration equals 12,000 Btu/hr.

$$1 \text{ ton} = 288,000 \text{ Btu}$$
$$= \frac{288,000 \text{ Btu}}{24 \text{ hr}}$$
$$= 12,000 \text{ Btu/hr}$$

On smaller jobs, such as dwellings, where there aren't detailed specifications, the electrician will need to ask the HVACR contractor about the size of the air conditioner to calculate the minimum branch circuit size. It is common for the contractor to specify the size of the unit in tons. This value isn't that helpful since it does not have a direct correlation to the sizing of the electrical branch circuit. The information needed to size the branch circuit can be found on the nameplate of the air conditioner or from the manufacturer's literature.

5.6.2 SI

Heat measurement in SI units is based on the calorie or the joule. A **calorie (cal)** is the amount of heat energy required to change the temperature of one gram of water by 1°C.

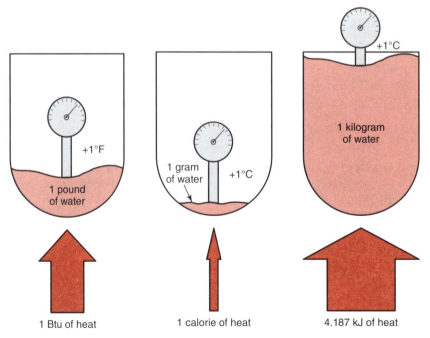

The **joule (J)** is a small unit of heat in the SI system. The amount of heat necessary to change the temperature of one kilogram of water by 1°C is 4,187 joules or 4.187 **kilojoules (kJ)**. Calories and joules are encountered when considering incident energy and arc flash boundaries. Electricians and designers use these values to determine what personal protective equipment is necessary to ensure worker safety when working around equipment with the potential to create an arc flash.

5.7 Pressure Measurement

Pressure is defined as force per unit of area: $P = F \div A$. Based on this equation, if the area remains the same, more force will result in greater pressure; less force will result in lower pressure. If the area is increased, the same amount of force will result in less pressure. If the area is decreased, the same amount of force will mean greater pressure.

Pressure exerted can be measured in pounds or newtons. The unit of area can be measured in square inches or square meters. A **square inch (in²)** is an area that is one inch by one inch. Likewise, a **square meter (m²)** is an area that is one meter by one meter.

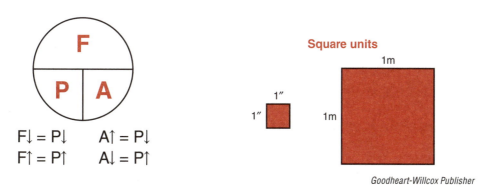

Goodheart-Willcox Publisher

Vacuum is the absence of atmospheric pressure. Vacuum can be measured in negative psi, but this is seldom done, as it is more accurately measured in inches of mercury (in. Hg) or microns (μm).

5.7.1 US Customary

The most common pressure measurement used in the United States is expressed in **pounds per square inch (psi)**. It is the pounds of force applied on a given area. For very small pressure levels and vacuum, a measurement in **inches of water column (in. WC)** may be used. This is also sometimes called *inches of water gauge (in. WG)*.

$$1 \text{ in. WC} = 0.036 \text{ psi}$$

$$1 \text{ psi} = 27.7 \text{ in. WC}$$

5.7.2 SI

In SI units, pressure is measured in pascals (Pa). A **pascal (Pa)** is the amount of force of one newton pushing on one square meter of area. Since a pascal is a small unit, it is common to use the **kilopascal (kPa)** as a unit of pressure. The kilopascal is equal to 1,000 pascals.

Vacuum measurement in the SI system uses the **micron (μm)**, which is 1/1,000 of a millimeter or 1/1,000,000 of a meter.

5.8 Air Velocity and Volume Measurement

Air velocity, or speed, is a measurement of the distance the air moves in a specific length of time. **Air volume** is the amount of air being moved over a specific length of time.

In some parts of the country, electricians supply and install residential bathroom fans. One of the ways bathroom fans are rated is by the air volume they move in cubic feet per minute. The size of the bathroom will be used to determine the minimum size of the fan in cubic feet per minute.

5.8.1 US Customary

In the US Customary system, air velocity is measured in **feet per minute (fpm)**, and air volume is measured in **cubic feet per minute (cfm)**.

5.8.2 SI

The SI unit used for air velocity is **meter per minute (m/min)**, and the unit for air volume is **cubic meter per minute (m³/min)**.

5.9 Measurement Conversions

It may be necessary to convert from US Customary units to SI units and vice versa. Many conversion factor lists are available in print format in pocket manuals and troubleshooting guides. Conversion apps are also available for electronic devices.

5.9.1 Linear Measurement Conversions

Linear Measurement Conversion

To convert from US Customary:	Multiply by:	To get SI:
inches	25.4	millimeters
inches	2.54	centimeters
inches	0.0254	meters
feet	304.8	millimeters
feet	30.48	centimeters
feet	0.3048	meters
feet	0.0003048	kilometers
To convert from SI:	**Multiply by:**	**To get US Customary:**
millimeters	0.03937	inches
centimeters	0.3937	inches
meters	39.37	inches
millimeters	0.003281	feet
centimeters	0.03281	feet
meters	3.281	feet
kilometers	3,281	feet

Goodheart-Willcox Publisher

Use of the chart shown allows for the conversion of varying units of measurement. For example, if 8″ needs to be converted to millimeters, the chart shows to multiply the 8″ by 25.4 to determine the millimeter equivalent.

$$8″ \times 25.40 = 203.20 \text{ mm}$$

Thus, 8″ equals 203.2 mm. To convert millimeters to inches, the chart shows a multiplier of 0.03937.

$$203.20 \text{ mm} \times 0.03937 = 7.9999840″$$

Thus, 203.2 mm equals 7.9999″, which can be rounded up to 8″.

5.9.2 Capacity Measurement Conversions

Capacity Measurement Conversion

To convert from US Customary:	Multiply by:	To get SI:
fluid ounces	29.573	milliliters
fluid ounces	0.02957	liters
pints	0.4732	liters
quarts	0.9463	liters
gallons	3.785	liters
To convert from SI:	**Multiply by:**	**To get US Customary:**
liters	33.814	fluid ounces
liters	2.113	pints
liters	1.057	quarts
liters	0.2642	gallons

Goodheart-Willcox Publisher

For the conversion of gallons to liters, the chart shows a multiplier of 3.785. Therefore, to convert 3 gallons to liters, the calculation would be as shown.

$$3 \text{ gal} \times 3.785 = 11.355 \text{ l}$$

Thus, 3 gallons equals 11.355 liters. To convert liters to gallons, use the multiplier 0.2642.

$$11.355 \text{ l} \times 0.2642 = 2.99999 \text{ gal}$$

Thus, 11.355 liters equals 2.9999 gallons, which can be rounded up to 3 gallons.

5.9.3 Weight Measurement Conversions

Weight Measurement Conversion

To convert from US Customary:	Multiply by:	To get SI:
ounces	28.349	grams
ounces	0.028349	kilograms
pounds	453.59	grams
pounds	0.4536	kilograms
To convert from SI:	Multiply by:	To get US Customary:
grams	0.03527	ounces
grams	0.002205	pounds
kilograms	35.274	ounces
kilograms	2.2046	pounds

Goodheart-Willcox Publisher

For a conversion of ounces to grams, the conversion chart multiplier is 28.349. To convert 12 ounces to grams, the calculation would be as shown.

$$12 \text{ oz} \times 28.349 = 340.188 \text{ g}$$

Thus, 12 ounces equals 340.188 grams. To convert grams to ounces, use the multiplier 0.03527.

$$340.188 \text{ g} \times 0.03527 = 11.998 \text{ oz}$$

Thus, 340.188 grams equals 11.998 ounces, which can be rounded up to 12 ounces.

5.9.4 Temperature Measurement Conversions

To convert from Fahrenheit to Celsius, use the following formula. Remember to perform operations within the parentheses first.

$$(°F - 32) \times \frac{5}{9} = °C$$

For example, a temperature of 70°F can be converted to Celsius as shown.

$$70° - 32 = 38°F$$

$$38°F \times \frac{5}{9} = \frac{190}{9}$$

$$190 \div 9 = 21.1°C$$

Seeing that the remainder of 1 would continue on infintely in this division equation, the calculation can stop at one decimal place. Thus, 70°F equals 21.1°C.

To convert from Celsius to Fahrenheit, the formula is as shown.

$$(°C \times \frac{9}{5}) + 32 = °F$$

A temperature of 21.1°C would convert to Fahrenheit as shown.

$$21.1°C \times \frac{9}{5} = 37.98°C$$

$$21.1°C \times 1.8 = 37.98°C$$

$$37.98°C + 32 = 69.98°F$$

Thus, 21.1°C equals 69.98°F, which can be rounded up to 70°F.

The scientific scales of Kelvin and Rankine are seldom used in the trades, but they are commonly used in laboratory settings and may be encountered in ultra-low temperature applications. To convert from Fahrenheit to Rankine, add 460° to the Fahrenheit value. To convert from Celsius to Kelvin, add 273° to the Celsius value.

5.9.5 Heat Measurement Conversions

Heating and cooling systems in the United States may be rated in Btu per hour, therms, tons, horsepower, pounds per hour, calories, joules, or kilojoules. It is often necessary to convert from one unit to another.

Conversions may be within the same measurement system, such as converting Btu/hr to horsepower (hp), which are both units in the US Customary system. However, it may be necessary to convert from US Customary units to the SI system of units as well.

Conversion Factors for Heat Measurement of the US Customary System

Multiply:	by	To obtain:
boiler hp	34.5	pounds of steam/hour
boiler hp	33,475	Btu/hour output
boiler hp	33.5	Mbh (1,000 Btu/hr output)
pounds of steam (@ 212°F)	970	Btu/hr
therm	100,000	Btu/hr
tons of refrigeration	12,000	Btu/hr
Divide:	**by**	**To obtain:**
pounds of steam/hour	34.5	boiler hp
Btu/hour output	33,475	boiler hp
Mbh (1,000 Btu/hr output)	33.5	boiler hp
Btu/hr	970	pounds of steam (@ 212°F)
Btu/hr	100,000	therm
Btu/hr	12,000	tons of refrigeration

Goodheart-Willcox Publisher

As an example of the need for calculating conversions, consider a boiler that is rated at a capacity of 300 horsepower. To determine the capacity (in pounds of steam per hour) required for a 300 hp boiler, use a multiplier of 34.5 as shown on the chart. This will convert hp into pounds of steam/hour.

$$300 \text{ hp} \times 34.5 = 10,350 \text{ pounds of steam/hour}$$

Conversion Factors for Heat Measurement of the SI System

Multiply:	by	To obtain:
calories	4.184	joules
calories	0.004184	kilojoules
joules	0.23901	calories
joules	0.001	kilojoules
kilojoules	1,000	joules
kilojoules	239.00574	calories

Goodheart-Willcox Publisher

The SI system also has various units of heat that may need to be converted. To convert from calories to kilojoules, the chart shows a multiplier of 0.004184. To convert 12,000,000 calories to kilojoules, the calculation would be as shown.

$$12,000,000 \text{ cal} \times 0.004184 = 50,208 \text{ kJ}$$

Thus, 12,000,000 calories equals 50,208 kilojoules. To convert kilojoules to calories, the chart shows a multiplier of 239.00574.

$$50,208 \text{ kJ} \times 239.00574 = 12,000,000 \text{ cal}$$

Thus, 50,208 kilojoules equals 12,000,000 calories.

Heat Measurement Conversion

To convert from US Customary:	Multiply by:	To get SI:
Btu	252.16	calories
Btu	1,055.06	joules
Btu	1.05506	kilojoules
To convert from SI:	**Multiply by:**	**To get US Customary:**
calories	0.0039685	Btu
joules	0.0009486	Btu
kilojoules	0.94782	Btu

Goodheart-Willcox Publisher

To convert from kilojoules to Btu, the conversion chart shows a multiplier of 0.94782. To convert 50,208 kilojoules to Btu, the calculation would be as shown.

$$50,208 \text{ kJ} \times 0.94782 = 47,588.146 \text{ Btu}$$

Thus, 50,208 kilojoules equals 47,588.146 Btu, which can be rounded to 47,588 Btu. To convert Btu to kilojoules, the chart shows a multiplier of 1.05506. Thus, to convert 47,588 Btu to kilojoules, the calculation would be as shown.

$$47,588 \text{ Btu} \times 1.05506 = 50,208.19 \text{ kJ}$$

Thus, 47,588 Btu equals 50,208.19 kilojoules, which can be rounded to 50,208 kilojoules.

5.9.6 Pressure Measurement Conversions

Conversion Factors for Pressure Measurements of the US Customary System

Multiply:	by	To obtain:
pounds per square inch (psi)	2.036	inches of mercury (in. Hg)
inches of mercury (in. Hg)	0.4912	pounds per square inch (psi)
atmosphere (atm)	14.7	pounds per square inch (psi)
atmosphere (atm)	29.92	inches of mercury (in. Hg)
inches of mercury (in. Hg)	0.00342	atmosphere (atm)

Goodheart-Willcox Publisher

To convert from pounds per square inch (psi) to inches of mercury (in. Hg), the conversion chart shows a multiplier of 2.036. Thus, to convert 40 pounds per square inch to inches of mercury, the calculation would be as shown.

$$40 \text{ psi} \times 2.036 = 81.44 \text{ in. Hg}$$

Thus, 40 pounds per squaire inch equals 81.44 inches of mercury. To convert from inches of mercury to pounds per square inch, the multiplier shown is 0.4912.

$$81.44 \text{ in. Hg} \times 0.4912 = 40.003328 \text{ psi}$$

Thus, 81.44 in. Hg equals 40.003328 psi, which can be rounded to 40 psi.

5

Conversion Factors for Pressure Measurements of the SI System

Multiply:	by	To obtain:
bar	100,000	pascal (Pa)
bar	100	kilopascal (kPa)
bar	1,000	millibar (mb)
bar	1.02	kilogram/centimeter² (kg/cm²)
kilopascal (kPa)	0.01	bar
kilopascal (kPa)	10	millibar (mb)
kilopascal (kPa)	1,000	pascal (Pa)
kilogram/centimeter² (kg/cm²)	980.7	millibar (mb)
kilogram/centimeter² (kg/cm²)	0.9807	bar
kilogram/centimeter² (kg/cm²)	98.07	kilopascal (kPa)
kilogram/centimeter² (kg/cm²)	735.6	millimeter of mercury (torr)
millibar (mb)	0.001	bar
millibar (mb)	0.1	kilopascal (kPa)
millibar (mb)	100	pascal (Pa)
millibar (mb)	0.00102	kilogram/centimeter² (kg/cm²)
millibar (mb)	0.7501	millimeter of mercury (torr)
millimeter of mercury (torr)	1.333	millibar (mb)
millimeter of mercury (torr)	133.3	pascal (Pa)
millimeter of mercury (torr)	0.1333	kilopascal (kPa)
millimeter of mercury (torr)	0.00136	kilogram/centimeter² (kg/cm²)
pascal (Pa)	1	newton/square meter (N/m²)
pascal (Pa)	0.001	kilopascal (kPa)
pascal (Pa)	0.0000102	kilogram/centimeter² (kg/cm²)
pascal (Pa)	0.00001	bar
pascal (Pa)	0.01	millibar (mb)
pascal (Pa)	0.007501	millimeter of mercury (torr)

Goodheart-Willcox Publisher

To convert millibar (mb) to kilogram/centimeter squared (kg/cm²), the conversion chart shows a multiplier of 0.00102. To convert 75 millibars to kilograms/centimeters squared, the calculation would be as shown.

$$75 \text{ mb} \times 0.00102 = 0.07650 \text{ kg/cm}^2$$

Thus, 75 millibars equals 0.0765 kilograms/centimeters squared. To convert from kilograms/centimeters squared to millibars, the chart shows a multiplier of 980.7.

$$0.0765 \text{ kg/cm}^2 \times 980.7 = 75.02355 \text{ mb}$$

Thus, 0.0765 kilograms/centimeters squared equals 75.02355 millibars, which can be rounded to 75 millibars.

Pressure Measurement Conversion

To convert from US Customary:	Multiply by:	To get SI:
pounds per square inch (psi)	6,895	pascal (Pa)
pounds per square inch (psi)	6.895	kilopascal (kPa)
pounds per square inch (psi)	0.07031	kilogram/centimeter^2 (kg/cm^2)
pounds per square inch (psi)	0.06895	bar
pounds per square inch (psi)	51.71	millimeters of mercury (torr)
pounds per square inch (psi)	6,895	newton/square meter (N/m^2)
To convert from SI:	**Multiply by:**	**To get US Customary:**
bar	14.5	pounds per square inch (psi)
bar	29.53	inches of mercury (in. Hg)
pascal (Pa)	0.000145	pounds per square inch (psi)
kilopascal (kPa)	0.145	pounds per square inch (psi)
kilogram/centimeter^2 (kg/cm^2)	14.22	pounds per square inch (psi)
millimeters of mercury (torr)	0.01934	pounds per square inch (psi)

Goodheart-Willcox Publisher

To convert from pounds per square inch to kilograms/centimeters squared, the conversion chart shows a multiplier of 0.07031. A conversion of 85 pounds per square inch to kilograms/centimeters squared would be calculated as shown here.

$$85 \text{ psi} \times 0.07031 = 5.97635 \text{ kg/cm}^2$$

Thus, 85 pounds per square inch equals 5.97635 kilograms/centimeters squared, which could be rounded to 5.98 kilograms/centimeters squared or 6 kilograms/centimeters squared. To convert 5.98 kilograms/centimeters squared to psi, use the multiplier 14.22.

$$5.98 \text{ kg/cm}^2 \times 14.22 = 85.0356 \text{ psi}$$

Thus, 5.98 kilograms/centimeters squared equals 85.0356 pounds per square inch, which can be rounded off to 85 pounds per square inch.

5.9.7 Velocity Measurement Conversions

Velocity Measurement Conversion

To convert from US Customary:	Multiply by:	To get SI:
feet per minute (fpm)	0.3048	meters per minute (m/min)
cubic feet per minute (cfm)	0.02832	cubic meters per minute (m³/min)
To convert from SI:	**Multiply by:**	**To get US Customary:**
meters per minute (m/min)	3.281	feet per minute (fpm)
cubic meters per minute (m³/min)	35.31	cubic feet per minute (cfm)

Goodheart-Willcox Publisher

To convert cubic feet per minute (cfm) to cubic meters per minute (m³/min), the conversion chart shows a multiplier of 0.02832. For a conversion of 500 cubic feet per minute to cubic meters per minute, the calculation would be as shown.

$$500 \text{ cfm} \times 0.02832 = 14.16 \text{ m}^3/\text{min}$$

Thus, 500 cubic feet per minute equals 14.16 cubic meters per minute. To convert cubic meters per minute to cubic feet per minute, use the multiplier 35.31.

$$14.16 \text{ m}^3/\text{min} \times 35.31 = 499.9896 \text{ cfm}$$

Thus, 14.16 cubic meters per minute equals 499.9896 cubic feet per minute, which can be rounded to 500 cubic feet per minute.

Review Questions

Name _____ Date _____ Class _____

Linear Measurement Conversion Exercises

Exercise 5-1

Convert the following into inches.

1. $12' =$ _____
2. $6' =$ _____
3. $21' =$ _____
4. $4'\text{-}1'' =$ _____
5. $11'\text{-}4'' =$ _____

6. $10'\text{-}1'' =$ _____
7. $13'\text{-}7'' =$ _____
8. $5'\text{-}9'' =$ _____
9. $22'\text{-}4'' =$ _____
10. $9'\text{-}2'' =$ _____

Exercise 5-2

Convert the following into the equivalent measurement in feet or feet and inches.

1. $24'' =$ _____
2. $120'' =$ _____
3. $180'' =$ _____
4. $77'' =$ _____
5. $89'' =$ _____

6. $149'' =$ _____
7. $63'' =$ _____
8. $15'' =$ _____
9. $134'' =$ _____
10. $92'' =$ _____

Adding and Subtracting Linear Measurements Exercises

Exercise 5-3

Add the following measurements. Express answers in feet and inches.

1. 1'-9" + 3'-3" = _____

2. 6'-5" + 4'-2" = _____

3. 12'-1" + 7'-6" = _____

4. 10'-6" + 4'-7" = _____

5. 8'-8" + 14'-9" = _____

6. 25'-4" + 10'-11" = _____

7. 13'-10" + 1'-4" = _____

8. 5'-3" + 4'-9" = _____

9. 2'-2" + 3'-5" + 4'-1" = _____

10. 11'-4" + 13'-5" + 26'-3" = _____

Exercise 5-4

Subtract the following measurements. Express answers in feet and inches.

1. 10'-9" – 6'-3" = _____

2. 8'-7" – 5'-1" = _____

3. 22'-7" – 12'-6" = _____

4. 15'-2" – 7'-7" = _____

5. 12'-8" – 11'-5" = _____

6. 25'-0" – 3'-10" = _____

7. 23'-1" – 5'-2" = _____

8. 2'-10" – 1'-11" = _____

9. 30'-2" – 7'-8" = _____

10. 13'-5" – 2'-8" = _____

Name _____   **Date** _____   **Class** _____

Multiplying and Dividing Linear Measurements Exercises

Exercise 5-5

Multiply the following measurements. Express answers in feet and inches.

1. $2'\text{-}6'' \times 2 =$ _____

2. $1'\text{-}2'' \times 6 =$ _____

3. $3'\text{-}8'' \times 10 =$ _____

4. $12'\text{-}2'' \times 3 =$ _____

5. $6'\text{-}4'' \times 7 =$ _____

6. $5'\text{-}9'' \times 4 =$ _____

7. $4'\text{-}3'' \times 5 =$ _____

8. $20'\text{-}1'' \times 24 =$ _____

9. $1'\text{-}6\ 1/4'' \times 4 =$ _____

10. $2'\text{-}6\ 1/2'' \times 3 =$ _____

Exercise 5-6

Divide the following measurements. Express answers in feet and inches.

1. $4'\text{-}6'' \div 2 =$ _____

2. $10'\text{-}8'' \div 8 =$ _____

3. $5'\text{-}9'' \div 3 =$ _____

4. $10'\text{-}10'' \div 5 =$ _____

5. $4'\text{-}8'' \div 7 =$ _____

6. $8'\text{-}4'' \div 20 =$ _____

7. $1'\text{-}10'' \div 11 =$ _____

8. $12'\text{-}6'' \div 10 =$ _____

9. $10'\text{-}2'' \div 4 =$ _____

10. $16'\text{-}2'' \div 8 =$ _____

Conversion Exercise

Exercise 5-7

Refer to the various conversion charts in this chapter to fill in the blanks. On the first blank of each question, write the multiplier. On the second blank, write the product of the original value times the multiplier.

1. 27 feet   _____ = _____ meters

2. 1,410 centimeters   _____ = _____ inches

3. 110 meters   _____ = _____ feet

4. 47 fluid ounces   _____ = _____ milliliters

5. 11.5 pints   _____ = _____ liters

6. 83.1 liters   _____ = _____ gallons

7. 473 ounces   _____ = _____ kilograms

8. 12 pounds   _____ = _____ grams

9. 124.3 kilograms   _____ = _____ pounds

10. 64°F   _____ = _____ °C

11. 92.4°F   _____ = _____ °C

12. –40°C   _____ = _____ °F

13. 950 boiler hp   _____ = _____ pounds of steam/hr

14. 24,000 pounds of steam (@212°F)   _____ = _____ Btu/hr

15. 1,200,000 joules   _____ = _____ kilojoules

16. 15.9 kilojoules   _____ = _____ calories

17. 345 Btu   _____ = _____ calories

18. 157,000 kilojoules   _____ = _____ Btu

19. –16.5 psi   _____ = _____ in. Hg

20. 147 kilopascal   _____ = _____ millibar

21. 1,250 pascal   _____ = _____ kilogram/cm^2

22. 14.29 kg/cm^2 _____ = _____ psi

23. 462 psi   _____ = _____ bar

24. 724 feet per minute   _____ = _____ meters per minute

25. 235 cubic meters per minute   _____ = _____ cubic feet per minute

CHAPTER 6

Algebraic Functions

Objectives

Information in this chapter will enable you to:

- Explain the various algebraic terms in the different formulas used in the electrical industry.
- Calculate equations while correctly applying the following algebraic properties: associative, distributive, and communicative.
- Determine the mean, median, and mode values of a set of numbers.
- Calculate complex equations in the proper order of operation, following the PEMDAS acronym.
- Solve for unknown values by manipulating equations to isolate the unknown.

Technical Terms

addends	commutative property	equation	mode
algebra		formula	parentheses ()
associative property	distributive property	mean	unknown
		median	variable

6.1 Introduction to Algebra

Algebra is a mathematical language used to calculate numerical values in situations where arithmetic alone will not suffice. The use of algebra includes working with unknown values, equations, and formulas. A basic understanding of algebra is necessary in the electrical industry. An understanding of the terminology used in algebra is essential in order to learn how to complete any algebraic calculations.

6.1.1 Algebraic Terms

A **formula** is an expression of a mathematical procedure, establishing a method of completing the calculation. For example, in the determination of the area of a rectangle, the formula is area equals length times width, which can be written as $A = l \times w$. The formula establishes the mathematics required to calculate the answer.

An **equation** is a comparison between two or more formulas or values. An equation is signified by the use of the equals sign (=) between the formulas or values. The equals sign requires that both sides of the equation have the same value.

For example, the simple equation $4 = 2 + 2$ shows that both sides of the equals sign are equal, as $2 + 2$ equals 4. Equations are commonly used to solve for unknown quantities, such as $X = 2 + 2$. In this case, of course, X equals 4.

An **unknown** or a **variable** is an unknown value or quantity that is designated by a letter in an equation or formula. The letter X is most commonly used, but any letter may be substituted for an unknown. The term *variable* is also used to denote the unknown value.

Parentheses () are used in formulas and equations to set apart a certain set of numbers or to indicate which portion of the calculation shall be performed first. For example, a formula may be written as $3(2 + 4) = X$. The parentheses indicate that the calculation of $2 + 4$ be performed before the multiplication.

The **associative property** allows numbers to be added and multiplied regardless of how they are grouped in an equation. In the equation $X = 3 + (2 + 4)$, the addition inside the parentheses would normally be performed first, thus:

$$X = 3 + 6 = 9$$

If the parentheses were moved to bracket the 3 and the 2, then:

$$X = (3 + 2) + 4$$

The answer would remain the same.

$$X = 5 + 4 = 9$$

This example illustrates the associative property of addition. Multiplication is also associative. In an equation with all addition or an equation with all multiplication, changing the order of operations will not change the answer.

The **distributive property** means that multiplication of a sum can be accomplished by multiplying each of the **addends**, or numbers that are added together, and then adding the products together. For example, in the equation $X = 3(2 + 4)$, the addition inside the parentheses would normally be performed first as shown.

$$X = 3(6) = 18$$

But the distributive property allows the multiplication to be performed first, meaning 3 would be multiplied by 2, and 3 would also be multiplied by 4.

$$X = ([3 \times 2] + [3 \times 4])$$

Those two products would then be added together, and the answer would remain the same as if calculated by adding the two values before multiplying.

$$X = (6 + 12) = 18$$

The answer is the same as if calculated the earlier way because multiplication is *distributive*.

The **commutative property** allows for the operations of multiplication and addition to be completed in any order. The sums or products will be the same no matter the order of values being multiplied. For example, take 6 + 4.

$$6 + 4 = 10$$

$$4 + 6 = 10$$

The answers are the same. For multiplication, the same rule applies.

$$6 \times 4 = 24$$

$$4 \times 6 = 24$$

Either way, the product is 24.

6.1.2 Mean, Median, and Mode Values

The terms mean, median, and mode are terms used when working with a group of values. **Mode** is the value in a series of values that is seen most often. **Mean** is the average of the values. Like any average, mean represents the sum of the values divided by the number of the values in the group.

$$X = \frac{A + B + C}{N}$$

where

N = the number of values being added

Median is the midpoint or the middle of the group of values, with half of the values being lower than the median and half of the values being higher than the median.

For example, the hourly electricity usage for a home was recorded for a 12-hour period. The list of readings shows a variance in measured usage in kilowatt-hours (kWh).

Daytime Utility Meter Reading

Time	kWh	Time	kWh
8 a.m.	12 kWh	2 p.m.	5 kWh
9 a.m.	11 kWh	3 p.m.	5 kWh
10 a.m.	6 kWh	4 p.m.	3 kWh
11 a.m.	5 kWh	5 p.m.	10 kWh
12 p.m.	5 kWh	6 p.m.	15 kWh
1 p.m.	4 kWh	8 p.m.	9 kWh

Goodheart-Willcox Publisher

6

The mean electricity usage for the 24-hour period shown would be an average of all of the readings. Thus, the sum of the readings is divided by the number of readings.

$$\text{Mean electricity usage} = \frac{\text{sum of all usage readings}}{\text{number of readings}}$$

$$= \frac{90}{12}$$

$$= 7.5 \text{ kWh}$$

The median temperature (the middle of all of the readings) would require the list to be rearranged in numerical order by temperature, rather than time.

Daytime Utility Meter Reading

Time	kWh	Time	kWh
4 p.m.	3 kWh	10 a.m.	6 kWh
1 p.m.	4 kWh	8 p.m.	9 kWh
11 a.m.	5 kWh	5 p.m.	10 kWh
12 p.m.	5 kWh	9 a.m.	11 kWh
2 p.m.	5 kWh	8 a.m.	12 kWh
3 p.m.	5 kWh	6 p.m.	15 kWh

Goodheart-Willcox Publisher

The median of the list is the middle, and because this list has an even number of values, the middle is between two of those values. Six of the values will be lower than the median, and six of the values will be higher than the median. The median, therefore, will be between 5 kWh and 6 kWh. When the number of values is even, the median is the average of the two values closest to the middle. Therefore, the median value is 5.5 kWh. In situations where there are an odd number of values, the median will simply be the value in the middle of the list. In this list, the mode, the value seen most often, is 5 kWh.

6.2 Working with Formulas and Equations

In solving equations, the goal is often to simplify the expression on one side of the equals sign in order to find the value of an unknown variable on the other side of the equals sign. This may involve more than one mathematical operation. In order to calculate the correct value, you must complete the operations in the proper order.

6.2.1 Order of Operations

When completing a calculation involving multiple operations, the order of operations is the sequence in which mathematical operations are performed. This is critical to finding the correct answer. While addition and multiplication have properties that allow the numbers to be added or multiplied in any order, subtraction and division do *not* have those same properties.

The acronym **PEMDAS** can be used to help remember the proper order for completing multiple operations in formulas and equations.

- **P**arentheses—Any operations inside parentheses are completed first.
- **E**xponents—Any numbers containing exponents are completed second.
- **M**ultiplication/**D**ivision—Multiplication and division are completed in left to right order.
- **A**ddition/**S**ubtraction—Addition and subtraction are completed in left to right order.

6.2.2 Solving for Unknown Values

Most electrical calculations involve solving for a single unknown variable. In the simplest form, solving for the unknown may require only arithmetic, such as the addition of a column of numbers.

For example, if an electrician needed to know the number of duplex receptacles that is required for a new home, a list of the number of receptacles installed in each room may appear as shown.

	Room	Quantity of Duplex Receptacles
A.	Kitchen	12
B.	Living room	7
C.	Bedrooms	16
D.	Den	7
E.	Bathrooms	4
F.	Hall/Foyer	3
G.	Misc.	3
	Total	52

Goodheart-Willcox Publisher

Since the total number of receptacles can be found by adding all the individual values together, a written formula or equation may not be necessary. But if an equation for this calculation was written, it would be as follows:

$$X = A + B + C + D + E + F + G$$

$$X = 12 + 7 + 16 + 7 + 4 + 3 + 3$$

$$X = 52$$

Calculations used in the electrical industry often involve solving equations for an unknown value. Most of the time, a formula can be used where values can simply be inserted into the equation to solve for the unknown. Watt's law, which is a commonly used electrical formula, is an example of this. The Watt's law power formula is $P = I \times E$. To solve for power (P) the circuit current (I) is multiplied by the applied voltage (E). For example, if the circuit current is 10 amperes and the applied voltage is 120 volts, the equation would be as shown.

$$P = 10 \times 120$$

$$P = 1,200 \text{ watts}$$

6.2.3 Rearranging Formulas

There are many formulas used in the electrical industry. They commonly involve addition, subtraction, multiplication, division, square roots, and squares. Each formula has many variations to allow a person to solve for the unknown value. It isn't necessary to memorize every variation of each formula; instead, they can be rearranged to solve for the unknown.

When rearranging formulas, variables and values are moved around to get the unknown value on one side of the equation by itself. It is important to remember that whatever you do to one side of the equation, you must also do to the other side. The following sections will explain and provide examples for rearranging common electrical formulas.

Addition and Subtraction

Formulas that contain addition and subtraction are rearranged by adding or subtracting variables until the unknown is by itself. An example of an electrical formula that contains addition is the formula for total resistance in a series circuit, $R_T = R_1 + R_2 + R_3.$ When solving for total resistance, the formula can be used as it is but to solve for one of the individual resistors, it must be rearranged to get the unknown value by itself on one side of the equation. To solve for R_3, $R_1 + R_2$ must be subtracted from both sides of the equation.

$$R_T = R_1 + R_2 + R_3$$
$$R_T - R_1 - R_2 = \cancel{R_1} + \cancel{R_2} + R_3$$
$$-R_1 \quad -R_2$$
$$R_3 = R_T - R_1 - R_2$$

Goodheart-Willcox Publisher

To convert $R_3 = R_T - R_1 - R_2$ back to the original equation, which is solving for total resistance, R_1 and R_2 are added to both sides of the equation.

$$R_3 = R_T - R_1 - R_2$$
$$R_3 + R_1 - R_2 = R_T - \cancel{R_1} - \cancel{R_2}$$
$$+R_1 \quad +R_2$$
$$R_T = R_1 + R_2 + R_3$$

Goodheart-Willcox Publisher

Multiplication and Division

Formulas that involve multiplication and division are rearranged by multiplying and dividing the variables to get the unknown value by itself. Ohm's law is an example, $E = I \times R$. In this example, E represents voltage, I represents current, and R represents resistance. In order to solve for I rather than E, the formula must be rearranged so that I is by itself. To accomplish this, both sides of the equation are divided by R.

$$E = I \times R$$

$$\frac{E}{R} = \frac{I \times R}{R}$$

$$\frac{E}{R} = \frac{I \times \cancel{R}}{\cancel{R}}$$

$$I = \frac{E}{R}$$

Goodheart-Willcox Publisher

To convert $I = E/R$ back to the original equation and solve for E, both sides must be multiplied by R.

$$I = \frac{E}{R}$$

$$I \times R = \frac{E}{\cancel{R}} \times \cancel{R}$$

$$E = I \times R$$

Goodheart-Willcox Publisher

Square and Square Roots

Formulas that involve squares and square roots are rearranged by either squaring or taking the square root of the variables to get the unknown value by itself. The formula $P = I^2 \times R$ is an example. To solve for I, the square must be eliminated. This is done by taking the square root of both sides of the equation. This equation will have an extra step, as R should be moved to the other side of the equals sign before applying the square root.

$$P = I^2 \times R$$

$$\frac{P}{R} = \frac{I^2 \times \cancel{R}}{\cancel{R}}$$

$$\frac{P}{R} = I^2$$

$$\sqrt{\frac{P}{R}} = \cancel{\sqrt{I^2}}$$

$$I = \sqrt{\frac{P}{R}}$$

Goodheart-Willcox Publisher

To convert $I = \sqrt{(P/R)}$ back to the original equation and solve for P, both sides must be squared.

$$I = \sqrt{\dfrac{P}{R}}$$

$$I^2 = \left(\sqrt{\dfrac{P}{R}}\right)^{\!\!\cancel{2}}$$

$$I^2 = \dfrac{P}{R}$$

$$I^2 \times R = \dfrac{P \times \cancel{R}}{\cancel{R}}$$

$$P = I^2 \times R$$

Goodheart-Willcox Publisher

Review Questions

Name _____ **Date** _____ **Class** _____

Mean and Median Values Exercise

Exercise 6-1

Find the mean and median values for each of the following sets of numbers.
Round answers to the nearest tenth.

1. 12, 27, 11, 41, 87, 1, 5

    Mean: _____

    Median: _____

2. 42, 99, 12, 3, 7

    Mean: _____

    Median: _____

3. 15, 6, 7, 111, 73, 4, 5

    Mean: _____

    Median: _____

4. 582, 345, 213, 12, 89

    Mean: _____

    Median: _____

5. 77, 1, 2, 345, 54, 6

    Mean: _____

    Median: _____

6. 66, 77, 88, 99

    Mean: _____

    Median: _____

Unknown Value Exercise

Exercise 6-2

Solve the following equations by finding the unknown value. Round answers to the nearest tenth.

1. $X = 3(47 + 2)$

   _____

2. $3Y = 9 \times 3 + 9$

   _____

3. $Z = 24 \times 7 \times 3 - 12$

   _____

4. $X = (23 \times 2) + (67 \times 11)$

   _____

5. $2Y = 65 - 33 + 3 \times 12$

   _____

6. $50Z = 42 \div 2 \times 7 + 3$

   _____

7. $X = (63 \div 21) \times (47 + 13) + (12 \div 4) - 7$

   _____

8. $Y = 82 - 23 - 7 + 147$

   _____

9. $Z = (12 \times 30 + 2) \div (24 \div 6)$

   _____

10. $X = (48 \div 6 \times 12) - (96 \div 1)$

   _____

Name _____   **Date** _____   **Class** _____

Rearranging Formulas Exercises

Exercise 6-3

Rearrange the formulas so the bolded unknown value is by itself on the left side of the equals sign.

1. $P = \mathbf{I} \times E$

2. $I_T = I_1 + \mathbf{I_2} + I_3$

3. $R = \dfrac{E}{I}$

4. $X_L = 2\pi \mathbf{F}L$

5. $X_C = \dfrac{1}{2\pi F \mathbf{C}}$

6. $E = \sqrt{\mathbf{P} \times R}$

7. $R = \sqrt{\mathbf{Z}^2 - X_L^{\,2}}$

8. $Z = \dfrac{E_2}{V\mathbf{A}}$

9. $E_C = I_C \times \mathbf{X_C}$

10. $I_{Line} = \mathbf{I_{Phase}} \times \sqrt{3}$

(Continued)

11. $P = I^2 \times R$

12. $I_L = \sqrt{I_T^{\,2} - I_R^{\,2}}$

13. $C = \dfrac{1}{2\pi F X_C}$

14. $R_T = \dfrac{1}{\dfrac{1}{R_1} + \dfrac{1}{R_2} + \dfrac{1}{R_3}}$

15. $Z = \dfrac{1}{\sqrt{\left(\dfrac{1}{R}\right)^2 + \left(\dfrac{1}{X_L} - \dfrac{1}{X_C}\right)^2}}$

Exercise 6-4

Solve the following equations for the unknown value.

1. $E = I \times R$
   $I = 6$
   $R = 10$

2. $I = \dfrac{P}{E}$
   $P = 3{,}600$
   $E = 240$

3. $I_T = I_1 + I_2 + I_3$
   $I_1 = 5$
   $I_2 = 3$
   $I_3 = 12$

Name _____   **Date** _____   **Class** _____

4. $L = \dfrac{X_L}{2\pi F}$

$X_L = 20$

$\pi = 3.14$

$F = 60$

5. $P = I^2 \times R$

$P = 200$

$R = 50$

6. $E_T = E_1 + E_2 + E_3$

$E_T = 120$

$E_1 = 25$

$E_2 = 10$

7. $C^2 = A^2 + B^2$

$A = 10$

$B = 6$

8. $E_T = \sqrt{VA \times Z}$

$E_T = 240$

$VA = 250$

9. $I = \sqrt{\dfrac{P}{R}}$

$I = 20$

$P = 1,000$

10. $R = \dfrac{E^2}{P}$

$R = 100$

$E = 240$

(Continued)

11. $R_T = \dfrac{R_1 \times R_2}{R_1 + R_2}$

 $R_1 = 10$

 $R_2 = 20$

12. $P = I \times E$

 $P = 3{,}000$

 $I = 50$

13. $R = \dfrac{P}{I^2}$

 $R = 20$

 $P = 200$

14. $X_C = \dfrac{1}{2\pi FC}$

 $X_C = 25$

 $\pi = 3.14$

 $F = 60$

15. $C^2 = A^2 + B^2$

 $C = 100$

 $A = 65$

CHAPTER 7

Geometric Functions

Objectives

Information in this chapter will enable you to:

- Identify and understand lines, points, angles, and various geometric terms.
- Identify various shapes found in the construction industry.
- Understand some of the inherent properties associated with different geometric shapes.
- Calculate the perimeter for various polygons and the circumference for circles.
- Calculate the area of polygons and circles and work with square units.
- Calculate the volume of cubes and cylinders and work with cubed units.

Technical Terms

acute angle	diameter	perimeter	square
acute triangle	equilateral triangle	perpendicular	square feet (ft^2)
altitude	exterior angle	pi (π)	square unit
angle	geometry	point	straight angle
arc	interior angle	polygon	surface
base	isosceles triangle	radius	tangent
chord	line	rectangle	trapezoid
circle	obtuse angle	right angle	triangle
circumference	obtuse triangle	right triangle	vertex
cubic feet (ft^3)	parallelogram	scalene triangle	volume
cubic unit		solid	

7.1 Introduction to Geometry

Geometry is a branch of mathematics that deals with the measurement, properties, and relationships of points, lines, angles, surfaces, and solids. Electricians and estimators often encounter situations where they are dealing with geometric functions.

The area of individual rooms, as well as the entire building, is used to determine the minimum number of luminaires or electric heaters and for sizing the electric service. Volume calculations are used to determine the amount of concrete needed for slabs and light posts. All of these procedures are based on the principles of geometry. Trigonometry, dealing with the angles and sides of triangles and their relationships, is discussed in Chapter 8, *Trigonometric Functions*.

7.1.1 Geometric Terms and Shapes

A **line**, as used in geometry, is a narrow strip or border that divides or connects areas or objects. Lines in geometry normally have both a starting and an ending point. Lines can be either straight or curved.

Goodheart-Willcox Publisher

A **point** is a definite position on a line. The point may be at the end of the line or anywhere along the line. A point is often used for reference, such as Point A or Point B.

Goodheart-Willcox Publisher

An **angle** is the figure formed by the intersection of two straight lines, and it determines the distance between the lines as they diverge from a common point. Angles are measured in degrees and have specific names that describe the type of angle. An **acute angle** is an angle that measures more than 0° but less than 90°.

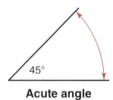

45°

Acute angle

Goodheart-Willcox Publisher

A **right angle** is formed by the intersection of two lines that form a 90° angle. Two lines forming a 90° angle are **perpendicular** to each other. All right angles measure 90°; thus, two right angles equal 180°, which is a straight line. A right angle is designated by placing a square at the intersection, showing the angle to be 90°.

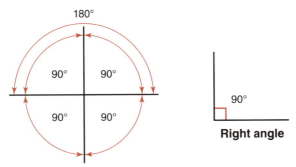

Goodheart-Willcox Publisher

An **obtuse angle** is an angle that measures more than 90° but less than 180°. A **straight angle** measures exactly 180°.

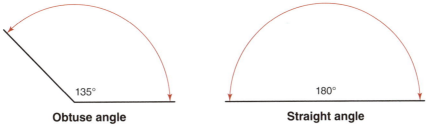

Goodheart-Willcox Publisher

In geometry, a **surface** is defined as a two-dimensional plane. An example is *surface area*, which is the surface of the outermost boundary of a three-dimensional object.

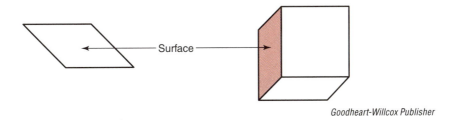

Goodheart-Willcox Publisher

A **polygon** is a two-dimensional closed plane with any number of straight sides. See examples of common polygons in the following table:

Common Polygons

Name	Triangle	Quadrilateral	Pentagon	Hexagon	Heptagon	Octagon
Number of Sides	3	4	5	6	7	8
Shape						

Goodheart-Willcox Publisher

A **square** is a polygon that has all four sides equal in length. A **rectangle** is a polygon with opposite sides equal in length and adjacent sides unequal in length.

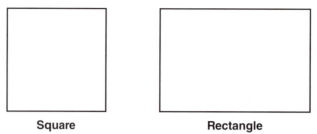

Square   Rectangle

A **triangle** is a polygon with three sides. The sides may or may not be equal in length. Triangles are classified by their angles, the sum of which always equals 180°.

An **interior angle** of a triangle is an angle inside the triangle. A triangle's **exterior angle** is angle between the extension of one side of the triangle and its adjacent side. The **vertex** is a corner of the triangle. All triangles have three vertices (plural for vertex).

The **base** of a triangle is any one of the sides of a triangle. The **altitude** is a perpendicular line from a triangle's base to the opposite vertex of the triangle. With some triangles, the base may need to be extended in order to draw the perpendicular line.

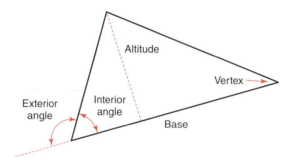

An **equilateral triangle** is a triangle with all three sides equal in length and all angles equal (thus 60° each). An **isosceles triangle** is a triangle with two equal sides. The angles opposite those sides are also equal. A **scalene triangle** is a triangle with no equal sides and no equal angles.

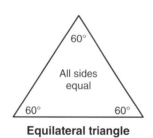
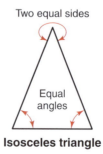
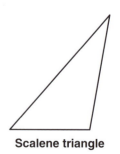

Equilateral triangle   **Isosceles triangle**   **Scalene triangle**

A **right triangle** is a triangle with one angle of 90°. An **acute triangle** is a triangle with all angles less than 90°. An **obtuse triangle** is a triangle with one angle that is more than 90°.

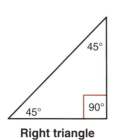

Right triangle

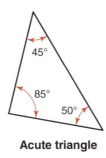

Acute triangle

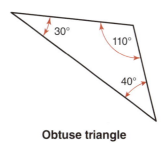

Obtuse triangle

Goodheart-Willcox Publisher

A **circle** is a closed curve with all points on the curve equidistant from the center. The distance around the outside of a circle is called the **circumference**. The distance from the center to an edge of the circle at any point is called the **radius**. The distance across the circle going through the center is called the **diameter**. The diameter is always twice the length of the radius.

A **chord** is a line that touches two points along the circumference of a circle without going through the center. A chord is always shorter than the diameter. A **tangent** is a straight line intersecting the circumference without entering the circle. An **arc** is a portion of the circumference.

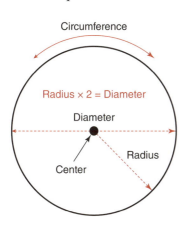

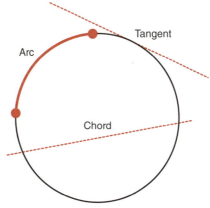

Goodheart-Willcox Publisher

In geometry, a **solid** is a three-dimensional (3D) object, such as a cube, cylinder, pyramid, or sphere. The measurement of the size of the content within a solid is called its *volume*. The face of the solid is called its *surface area*.

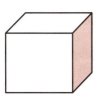

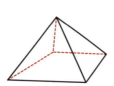

Goodheart-Willcox Publisher

7

7.2 Working with Geometric Shapes

Buildings may be any shape, and they are constructed of squares, rectangles, triangles, and even circles. For example, when trying to determine the area of a building, a variety of geometric shapes may be encountered within the building envelope. Electricians will have to install equipment, such as luminaires, and wiring methods, such as raceways, in conjunction with the geometric shapes of buildings. The engineer, installer, and electrician must be accustomed to working with a multitude of geometric shapes and the principles and formulas associated with those shapes.

7.2.1 Perimeter and Circumference

The **perimeter** of a shape is the distance around the outside of the shape. One example of where electricians and estimators use perimeter calculations is when determining the amount of conduit it will take to run around an object, room, or building.

Perimeter is determined by adding together the lengths of the sides of the shape. When all sides of an object have an equal length, time can be saved by multiplying the number of sides by their length to determine the perimeter. Since many shapes do not have sides that are equal in length, the length of each of the sides are added together. For example, if a room measures 16′ wide by 24′ long, the perimeter would be calculated as shown.

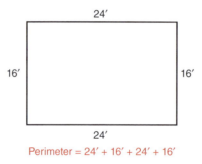

Perimeter = 24′ + 16′ + 24′ + 16′

Since this room is rectangular (having two walls the same size and two other walls the same size), its perimeter could be calculated another way as shown.

$$2 \text{ walls} \times 16' = 32'$$
$$\underline{+\ 2 \text{ walls} \times 24' = 48'}$$
$$80'$$

Regardless of the method used, the perimeter, or the distance around the outside of the room, is 80′.

The perimeter of a circle is called its *circumference*. It is determined by using either of the formulas shown.

$$C = \pi d \quad \text{or} \quad C = 2\pi r$$

where

C = circumference

π = pi (3.14159)

d = diameter

r = radius

The constant **pi (π)** is a mathematical determination of the relationship between the circumference and the diameter of a circle that never changes. It is the quotient of the circumference of the circle divided by the diameter. The number of pi has been mathematically determined to continue infinitely without repeating any sequence of digits, but it is typically shortened to 3.14 for calculation purposes.

$$\text{Circumference} \div \text{diameter} = \pi$$

For example, if the diameter of a round cooling tower is 16′, the circumference, or distance around the outside edge of the tower, would be calculated using the formula $C = \pi d$.

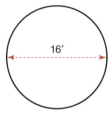

16′

Goodheart-Willcox Publisher

$$C = (3.14)16'$$

Placing the parentheses around the 3.14, with the 16 outside the parentheses, simply indicates that the two values are to be multiplied.

$$C = 50.24'$$

The circumference of a 16′ diameter cooling tower is 50.24′.

7.2.2 Area of Plane Surfaces

The area of a plane surface is the amount of space that is occupied or covered by that surface. Area is measured in **square units**, such as square inches (in^2) or square feet (ft^2). Electricians and designers must often calculate the area of rooms and buildings. A few examples include performing electric service calculations, determining the minimum lighting load, and calculating the number of light fixtures or electric heating units.

To calculate the area of a plane surface, its shape as well as its dimensions will be considered. For a rectangular area with two walls that are the same length and the other two walls the same length, the area is calculated by using the formula shown.

$$\text{Area } (A) = \text{length } (l) \times \text{width } (w)$$

This formula is used for any square or rectangular polygon. In a room that measures 16′ by 24′, the calculation of area would be as shown.

$$A = 16' \times 24'$$

$$= 384 \text{ ft}^2$$

Any number multiplied by itself is referred to as squared, so when a unit is multiplied by the same unit, it becomes that unit squared. In this example, feet are multiplied by feet, so the product is in **square feet (ft²)**. Thus, 16′ × 24′ equals 384 square feet. Square feet may be abbreviated as either sq ft or ft².

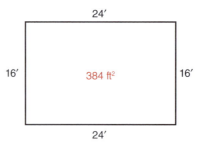

Goodheart-Willcox Publisher

A **parallelogram** is a four-sided figure with opposite sides that are parallel. Its area calculation is the same as for a rectangle or square, but with a parallelogram, the width is called *height*. A perpendicular line must be drawn from the base to obtain the correct measurement of height. Therefore, the formula is as shown.

$$\text{Area } (A) = \text{length } (l) \times \text{height } (h)$$

$$A = l \times h$$

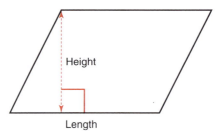

Goodheart-Willcox Publisher

For example, if the length is 40′ and the height is 20′, the area of the parallelogram would be 800 ft².

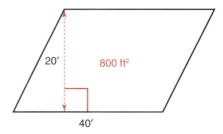

Goodheart-Willcox Publisher

$$40′ \times 20′ = 800 \text{ ft}^2$$

A **trapezoid** is another four-sided polygon, with two parallel sides and adjacent sides that may or may not be parallel to each other.

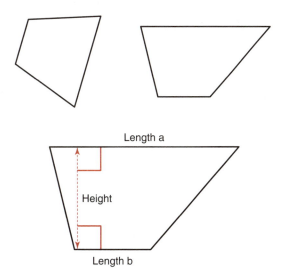

Goodheart-Willcox Publisher

To calculate the area of a trapezoid, the formula requires knowing the height of the trapezoid—the same as in a parallelogram. A line perpendicular to the parallel sides is drawn to establish the height measurement.

The length of the two parallel sides is then added together, and the height is multiplied by the combined length. One-half of that product will equal the area of the trapezoid. Therefore, the formula for the area of a trapezoid is as shown.

$$\text{Area } (A) = \frac{(\text{length } a + \text{length } b) \times \text{height}}{2}$$

$$A = \frac{(l_a + l_b) \times h}{2}$$

For example, if length a is 20′, length b is 12′, and the height is 14′, then the calculation would be 20′ plus 12′ times 14′, and that product divided by 2.

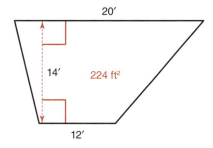

Goodheart-Willcox Publisher

$$A = \frac{(20' + 12') \times 14'}{2}$$

$$= \frac{(32') \times 14'}{2}$$

$$= \frac{448 \text{ ft}^2}{2}$$

$$= 224 \text{ ft}^2$$

Any triangle can be considered to be one-half of a rectangle, so the formula for the area of a triangle uses one half the length times the width. In triangles, however, the length is called the *base*, and the width is called *height*.

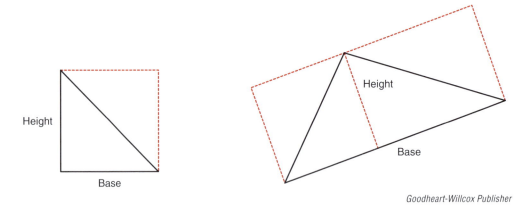

The base of a triangle is any one of the sides of the triangle, and the height (also called the *altitude*) is a perpendicular line from the base to the opposite angle of the triangle. With some triangles, the base may have to be extended in order to draw the perpendicular line. With any right triangle, the perpendicular line is already established at the right angle. The formula for calculating the area of a triangle is shown here.

$$A = \frac{b \times h}{2}$$

For example, if a triangle has a base that measures 20″, and a height that measures 12″, the calculation would be 20″ times 12″ divided by 2.

$$A = \frac{20'' \times 12''}{2}$$

$$= 120 \text{ in}^2$$

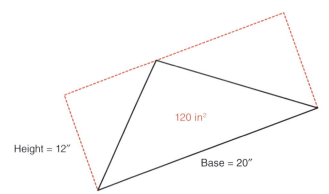

The area and volume of polygons and different shapes can be converted to different units using conversion factors, such as those shown in the following table.

Area and Volume Conversions

Multiply:	by	To Obtain:
square feet	144	square inches
square feet	0.111	square yards
square feet	0.092903	square meters
square inches	0.006944	square feet
square inches	645.16	square millimeters
square inches	0.0007716	square yards
square yards	9	square feet
cubic feet	1,728	cubic inches
cubic inches	16,387.064	cubic millimeters
cubic feet	0.028	cubic meters
cubic meters	35.315	cubic feet
cubic millimeters	0.000061	cubic inches
cubic millimeters	0.000000035	cubic feet
cubic yards	27	cubic feet
Divide:	**by**	**To Obtain:**
square inches	144	square feet
cubic inches	1,728	cubic feet

Goodheart-Willcox Publisher

7.2.3 Area of a Circle

The area of a circle can be calculated by using the formula pi times radius squared.

$$A = \pi r^2$$

For example, to calculate the area of the bottom of a circle that measures 16′ in diameter, you would need to find the radius. Since the radius is half the diameter, the radius of the circle would be 8′. The radius (8) is then squared, equaling 64.

$$A = \pi 8'^2$$

$$= 3.14(64 \text{ ft}^2)$$

$$= 200.96 \text{ ft}^2$$

Therefore, the area of a circle with a diameter of 16′ is 200.96 ft².

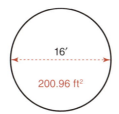

16′

200.96 ft²

Goodheart-Willcox Publisher

7.2.4 Volume

Volume can be defined as the amount of space that is occupied by a three-dimensional object. Because an object has three dimensions, volume is measured in **cubic units**, such as cubic inches (in^3) or cubic feet (ft^3).

Electricians will often have to pour concrete slabs and bases to support or conceal electrical equipment. The amount of concrete needed is determined by performing a volume calculation. Examples include light pole bases, transformer pads, duct banks, and solar rack foundations.

Volume of a Cube

The volume of a cube is calculated by using the following formula:

$$\text{Volume } (V) = \text{length } (l) \times \text{width } (w) \times \text{height } (h)$$

$$V = l \times w \times h$$

This formula is used for any rectangular three-dimensional object. To calculate volume, the three dimensions (length, width, and height) are multiplied by each other. For example, in a room that measures 16′ by 24′ by 8′, the calculation of area would be length times width, which results in 384 ft².

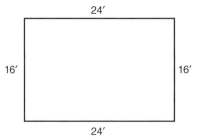

Goodheart-Willcox Publisher

$$A = 16' \times 24'$$

$$= 384 \text{ ft}^2$$

When the third dimension (height) is incorporated, the object takes on another aspect: volume.

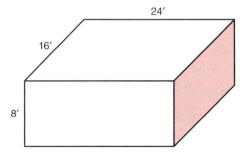

Goodheart-Willcox Publisher

Just as any number multiplied by itself is squared, any number that is multiplied by itself two times is cubed. When square feet is multiplied by feet, the product is in **cubic feet (ft³)**. For example, 16′ times 24′ equals 384 ft², and when multiplied by 8′, it equals 3,072 cubic feet. Cubic feet may be abbreviated as either cu ft or ft³.

$$V = l \times w \times h$$

$$= 16' \times 24' \times 8'$$

$$= 3{,}072 \text{ ft}^3$$

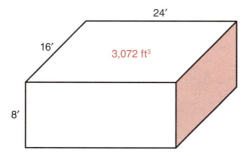

Goodheart-Willcox Publisher

Volume of a Cylinder

To calculate the volume of a cylinder, the radius or diameter of the cylinder must be known. Once the radius is known, the area of the end of the cylinder can be calculated using the following formula:

$$A = \pi r^2$$

For example, to calculate the area of the cylindrical concrete foundation for a solar panel rack that measures 3′ in diameter, the formula can be applied as follows. Because the diameter is known to be 3′ and the radius is 1/2 of the diameter, the radius of the foundation is 1.5′.

$$A = \pi(1.5'^2)$$

$$= 3.14(2.25 \text{ ft}^2)$$

$$= 7.065 \text{ ft}^2$$

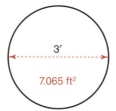

Goodheart-Willcox Publisher

The area of the foundation pictured, which measures 3′ in diameter, is 7.065 ft². The formula used for the volume of a cylinder will be volume equals area of the base times the length (or height) of the cylinder.

$$V = A \times l$$

$$= (\pi \times r^2) \times l$$

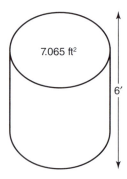

Continuing with the above example, if the cylinder has a diameter of 3′ and is 6′ long (or deep), the formula for volume would be as shown.

$$V = \pi \times r^2 \times l$$
$$= \pi \times 1.5'^2 \times 6'$$
$$= \pi \times 2.25 \text{ ft}^2 \times 6'$$
$$= 7.065 \text{ ft}^2 \times 6'$$
$$= 42.39 \text{ ft}^3$$

The volume of the foundation, therefore, is 42.39 ft³.

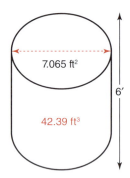

When volume calculations are made for concrete foundations, the number of cubic feet needs to be converted to cubic yards. Since one cubic yard is equal to 27 cubic feet, the amount of concrete in cubic yards is calculated by dividing the volume in cubic feet by 27.

$$\frac{42.39}{27} = 1.57 \text{ cubic yards}$$

Review Questions

Name _____ **Date** _____ **Class** _____

Perimeter and Circumference Exercises

Exercise 7-1

Calculate the perimeter of each of the following polygons.

1. _____

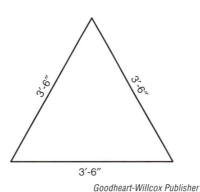

3'-6"   3'-6"

3'-6"

Goodheart-Willcox Publisher

2. _____

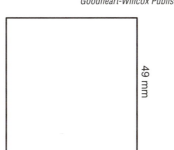

49 mm

49 mm

Goodheart-Willcox Publisher

3. _____

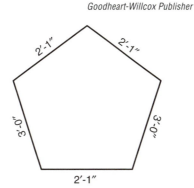

2'-1"   2'-1"

3'-0"   3'-0"

2'-1"

Goodheart-Willcox Publisher

4. _____

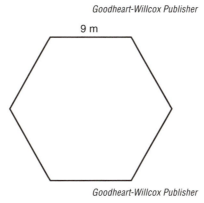

9 m

Goodheart-Willcox Publisher

(Continued)

5. _____

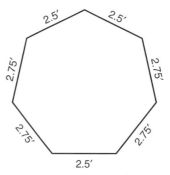

Goodheart-Willcox Publisher

6. _____

Goodheart-Willcox Publisher

7. _____

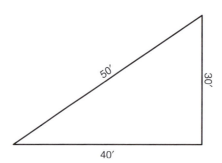

Goodheart-Willcox Publisher

8. _____

Goodheart-Willcox Publisher

Name _____   **Date** _____   **Class** _____

Exercise 7-2

Examine each circle in this exercise. Then calculate and record their three major values: circumference, diameter, and radius. Use 3.14 for pi, and round answers to the nearest hundredth.

1. Circumference: _____

   Diameter: _____

   Radius: _____

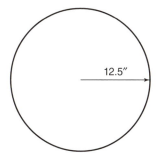

12.5″

Goodheart-Willcox Publisher

2. Circumference: _____

   Diameter: _____

   Radius: _____

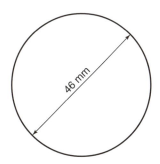

46 mm

Goodheart-Willcox Publisher

3. Circumference: _____

   Diameter: _____

   Radius: _____

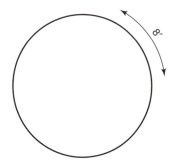

8′

Goodheart-Willcox Publisher

4. Circumference: _____

   Diameter: _____

   Radius: _____

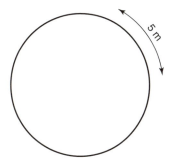

5 m

Goodheart-Willcox Publisher

Polygon Area Exercise

Exercise 7-3

Calculate the area of the following polygons. Round answers to the nearest hundredth.

1. Area: _____

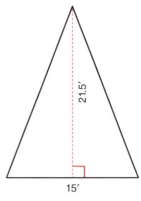

Goodheart-Willcox Publisher

2. Area: _____

Goodheart-Willcox Publisher

3. Area: _____

Goodheart-Willcox Publisher

4. Area: _____

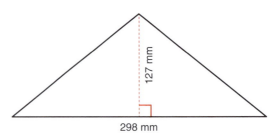

Goodheart-Willcox Publisher

Name _____   **Date** _____   **Class** _____

5. Area: _____

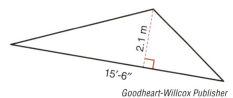

2.1 m

15'-6"

Goodheart-Willcox Publisher

6. Area: _____

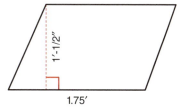

1'-1/2"

1.75'

Goodheart-Willcox Publisher

7. Area: _____

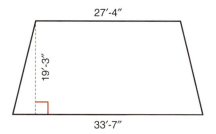

27'-4"

19'-3"

33'-7"

Goodheart-Willcox Publisher

8. Area: _____

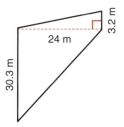

3.2 m

24 m

30.3 m

Goodheart-Willcox Publisher

Combined Area Exercises

Exercise 7-4

Calculate the area of the following shapes. Round answers to the nearest hundredth.

1. Area: _____

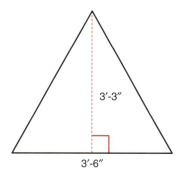

3'-3"

3'-6"

Goodheart-Willcox Publisher

(Continued)

2. Area: _____

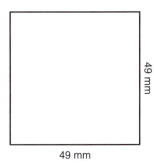

49 mm

49 mm

Goodheart-Willcox Publisher

3. Area: _____

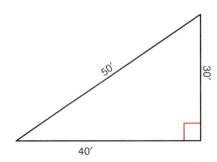

50'

30'

40'

Goodheart-Willcox Publisher

4. Area: _____

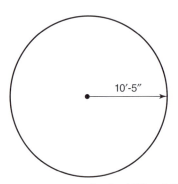

10'-5"

Goodheart-Willcox Publisher

5. Area: _____

144 mm

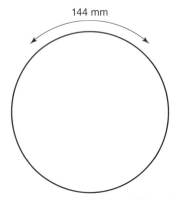

Goodheart-Willcox Publisher

Name _____   **Date** _____   **Class** _____

6. Area: _____

18'-6"

Goodheart-Willcox Publisher

Practical Exercise 7-5

Using the dimensions shown on the drawing provided, calculate the following space allocations. (Use the dimensions shown and ignore wall thicknesses when determining the area of rooms.) Round answers to the nearest hundredth.

1. What is the area of the Administrative Office?

   _____

2. How much area is allocated as a Waiting Room?

   _____

3. What is the area of the space dedicated to office cubicles?

   _____

4. What is the area of the Conference Room?

   _____

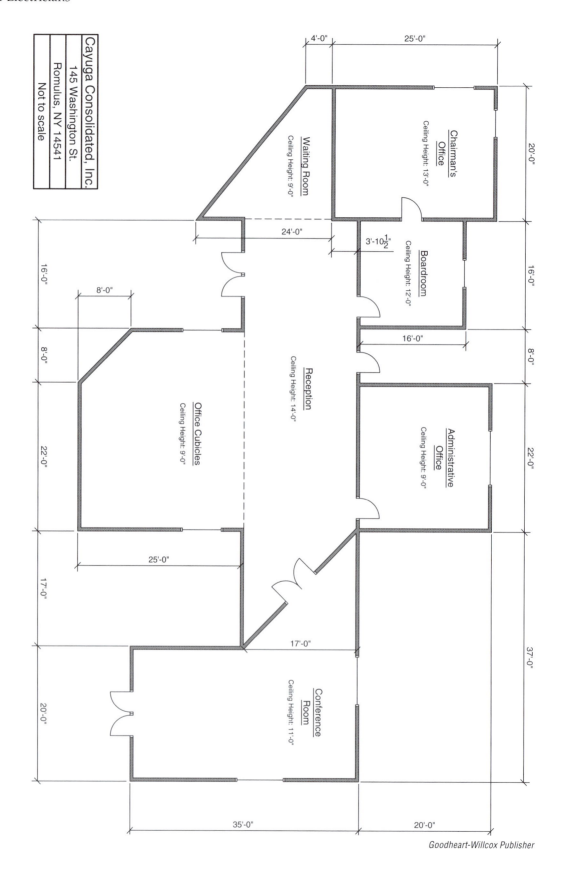

Cayuga Consolidated, Inc.
145 Washington St.
Romulus, NY 14541

Not to scale

Waiting Room
Ceiling Height: 9'-0"

Chairman's
Office
Ceiling Height: 13'-0"

Boardroom
Ceiling Height: 12'-0"

Reception
Ceiling Height: 14'-0"

Office Cubicles
Ceiling Height: 9'-0"

Administrative
Office
Ceiling Height: 9'-0"

Conference
Room
Ceiling Height: 11'-0"

4'-0" 25'-0" 20'-0"

24'-0" 3'-10½" 16'-0"

16'-0" 8'-0" 8'-0"

8'-0" 16'-0"

22'-0" 22'-0"

25'-0" 17'-0" 37'-0"

17'-0"

20'-0"

35'-0" 20'-0"

Goodheart-Willcox Publisher

Name _____   **Date** _____   **Class** _____

5. The chairman would like electric floor heat in his office. Determine the minimum amount of heat, in watts, that will be needed. Use 8 watts per square foot for this calculation.

_____

6. What is the total area of the entire building?

_____

The *National Electrical Code* calculates the general lighting load of a space by multiplying the area of the space by a specified value found in *Table 220.42(A)*. The value assigned to office spaces is 1.3 volt-amperes per square foot.

7. Determine the minimum general lighting load for Cayuga Consolidated, Inc.

_____

Cubic Volume Exercise

Exercise 7-6

Calculate the volume of the following cubes. Round answers to the nearest hundredth.

1. Volume: _____

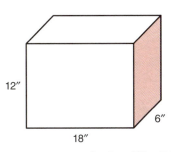

12″

6″

18″

Goodheart-Willcox Publisher

(Continued)

2. Volume: _____

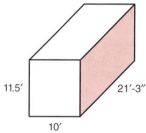

11.5'   21'-3"

10'

Goodheart-Willcox Publisher

3. Volume: _____

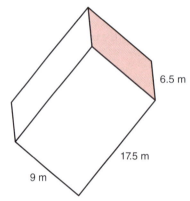

6.5 m

17.5 m

9 m

Goodheart-Willcox Publisher

4. Volume: _____

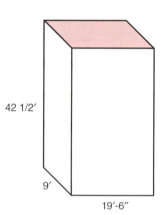

42 1/2'

9'

19'-6"

Goodheart-Willcox Publisher

Name _____   **Date** _____   **Class** _____

Cylindrical Volume Exercises

Exercise 7-7

Calculate the volume of each of the following cylinders. Round answers to the nearest hundredth.

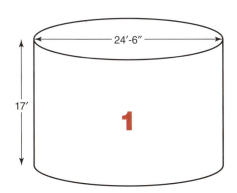

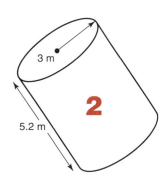

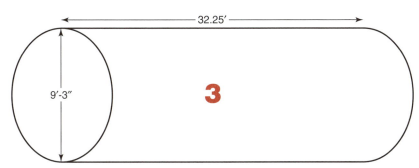

1. Calculate the volume for cylinder 1.

   _____

2. Calculate the volume for cylinder 2.

   _____

(Continued)

3. Calculate the volume for cylinder 3.

_____

Practical Exercise 7-8

A parking lot needs to have 20 light poles installed. Each light will sit on a concrete base that is a cylinder. The light bases will be 18″ in diameter, stick out of the ground 2′, and extend 6′ below the ground. Round answers to the nearest hundredth.

1. Calculate the amount of concrete, in cubic feet, that is required for one of the light pole bases.

_____

2. Calculate the total amount of concrete, in cubic feet, that is required for all the light pole bases.

_____

3. Convert the total amount of concrete required from cubic feet to cubic yards.

_____

CHAPTER 8

Trigonometric Functions

Objectives

Information in this chapter will enable you to:

- Recognize trigonometry as it relates to shapes and angles found in the electrical industry.
- Use sine, cosine, tangent, and a trigonometric chart or scientific calculator to determine unknown angles of a right triangle.
- Use sine, cosine, tangent, and a trigonometric chart or scientific calculator to determine unknown side lengths of a right triangle for a given angle.
- Calculate unknown lengths of a right triangle using the Pythagorean Theorem.

Technical Terms

angle theta (Angle θ)

cosecant

cosine (cos)

Pythagorean Theorem

sine (sin)

tangent (tan)

trigonometry (trig)

8.1 Introduction to Trigonometry

Trigonometry (trig) is defined as a branch of mathematics dealing with the angles and sides of triangles and their relationships. The electrical industry uses a lot of right triangles, also called *3-4-5 triangles*. Alternating current (ac) circuits containing inductors and capacitors use right triangles in their calculations. Multipliers used for bending conduit offsets and saddles are derived from right triangles. When determining if something is square to a wall or if a room is square, a 3-4-5 triangle is often used. Even though you may not think of the work electricians do as involving triangles, trigonometry has been used to create the multipliers and formulas that we take for granted.

8.1.1 Trigonometric Functions

The three trig functions you will use to solve right triangles are sine (sin), cosine (cos), and tangent (tan). Trig functions can seem confusing, but they just represent ratios. They are the ratio of one side of a triangle to another based on a particular angle. The angle used within the functions is typically referred to as **angle theta (Angle θ)**. The function to be used is determined by the information that is known about a triangle.

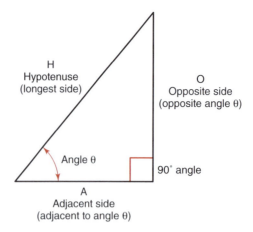

A right triangle will have a square in the corner of the triangle that makes a 90-degree angle. In the picture, the reference angle (Angle θ) is in the lower left corner. The opposite and adjacent sides of the triangle are determined by Angle θ. The opposite side is across from Angle θ, which is on the right side of the triangle. The adjacent side is adjacent to Angle θ, which is on the bottom. The hypotenuse is the longest side of the triangle and is always across from a 90-degree angle.

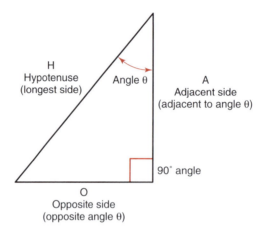

If Angle θ is moved to the top right of the triangle as pictured, the opposite and adjacent sides will also move. The side opposite from Angle θ is now on the bottom of the triangle, while the side adjacent to Angle θ is now on the right side. The hypotenuse did not change, as the 90-degree Angle is still on the bottom right of the triangle. For the remainder of the chapter, Angle θ will be in the bottom left corner of the triangles for consistency and to avoid confusion.

Sine (sin) is a number that represents the ratio of the length of the side opposite Angle θ divided by the length of the longest side (hypotenuse).

$$\text{Sin } \theta = \frac{\text{Opposite side}}{\text{Hypotenuse}}$$

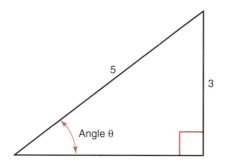

Goodheart-Willcox Publisher

For example, if the length of the opposite side is 3, and the length of the hypotenuse is 5, the calculation of the sine of Angle θ is 3 divided by 5, which equals 0.6. The formula would be written as shown.

$$\text{Sin } \theta = \frac{\text{Opposite side}}{\text{Hypotenuse}}$$

$$= \frac{3}{5}$$

$$= 0.6$$

By using a trigonometric chart, it can be determined that Angle θ is approximately 37°. A scientific calculator can also be used to find the angle by using the (sin⁻¹) button of 0.6 as shown in the image. The answer provided by the calculator is 36.87. The calculator provides an answer with a higher degree of accuracy because it isn't rounding to the nearest degree like the trigonometric chart.

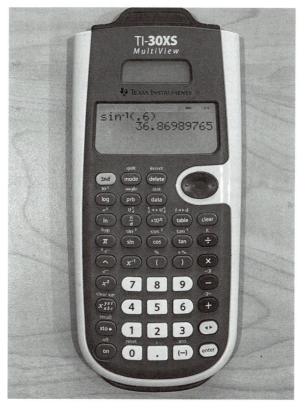

Goodheart-Willcox Publisher

Trigonometric Chart

Angle	Radian Measure	Sin	Cos	Tan	Angle	Radian Measure	Sin	Cos	Tan
0	0.00000	0.00000	1.00000	0.00000	46	0.80285	0.71934	0.69466	1.03553
1	0.01745	0.01745	0.99985	0.01746	47	0.82030	0.73135	0.68200	1.07237
2	0.03491	0.03490	0.99939	0.03492	48	0.83776	0.74314	0.66913	1.11061
3	0.05236	0.05234	0.99863	0.05241	49	0.85521	0.75471	0.65606	1.15037
4	0.06981	0.06976	0.99756	0.06993	50	0.87266	0.76604	0.64279	1.19175
5	0.08727	0.08716	0.99619	0.08749	51	0.89012	0.77715	0.62932	1.23490
6	0.10472	0.10453	0.99452	0.10510	52	0.90757	0.78801	0.61566	1.27994
7	0.12217	0.12187	0.99255	0.12278	53	0.92502	0.79864	0.60182	1.32704
8	0.13963	0.13917	0.99027	0.14054	54	0.94248	0.80902	0.58779	1.37638
9	0.15708	0.15643	0.98769	0.15838	55	0.95993	0.81915	0.57358	1.42815
10	0.17453	0.17365	0.98481	0.17633	56	0.97738	0.82904	0.55919	1.48256
11	0.19199	0.19081	0.98163	0.19438	57	0.99484	0.83867	0.54464	1.53986
12	0.20944	0.20791	0.97815	0.21256	58	1.01229	0.84805	0.52992	1.60033
13	0.22689	0.22495	0.97437	0.23087	59	1.02974	0.85717	0.51504	1.66428
14	0.24435	0.24192	0.97030	0.24933	60	1.04720	0.86603	0.50000	1.73205
15	0.26180	0.25882	0.96593	0.26795	61	1.06465	0.87462	0.48481	1.80405
16	0.27925	0.27564	0.96126	0.28675	62	1.08210	0.88295	0.46947	1.88073
17	0.29671	0.29237	0.95630	0.30573	63	1.09956	0.89101	0.45399	1.96261
18	0.31416	0.30902	0.95106	0.32492	64	1.11701	0.89879	0.43837	2.05030
19	0.33161	0.32557	0.94552	0.34433	65	1.13446	0.90631	0.42262	2.14451
20	0.34907	0.34202	0.93969	0.36397	66	1.15192	0.91355	0.40674	2.24604
21	0.36652	0.35837	0.93358	0.38386	67	1.16937	0.92050	0.39073	2.35585
22	0.38397	0.37461	0.92718	0.40403	68	1.18682	0.92718	0.37461	2.47509
23	0.40143	0.39073	0.92050	0.42447	69	1.20428	0.93358	0.35837	2.60509
24	0.41888	0.40674	0.91355	0.44523	70	1.22173	0.93969	0.34202	2.74748
25	0.43633	0.42262	0.90631	0.46631	71	1.23918	0.94552	0.32557	2.90421
26	0.45379	0.43837	0.89879	0.48773	72	1.25664	0.95106	0.30902	3.07768
27	0.47124	0.45399	0.89101	0.50953	73	1.27409	0.95630	0.29237	3.27085
28	0.48869	0.46947	0.88295	0.53171	74	1.29154	0.96126	0.27564	3.48741
29	0.50615	0.48481	0.87462	0.55431	75	1.30900	0.96593	0.25882	3.73205
30	0.52360	0.50000	0.86603	0.57735	76	1.32645	0.97030	0.24192	4.01078
31	0.54105	0.51504	0.85717	0.60086	77	1.34390	0.97437	0.22495	4.33148
32	0.55851	0.52992	0.84805	0.62487	78	1.36136	0.97815	0.20791	4.70463
33	0.57596	0.54464	0.83867	0.64941	79	1.37881	0.98163	0.19081	5.14455
34	0.59341	0.55919	0.82904	0.67451	80	1.39626	0.98481	0.17365	5.67128
35	0.61087	0.57358	0.81915	0.70021	81	1.41372	0.98769	0.15643	6.31375
36	0.62832	0.58779	0.80902	0.72654	82	1.43117	0.99027	0.13917	7.11537
37	0.64577	0.60182	0.79864	0.75355	83	1.44862	0.99255	0.12187	8.14435
38	0.66323	0.61566	0.78801	0.78129	84	1.46608	0.99452	0.10453	9.51436
39	0.68068	0.62932	0.77715	0.80978	85	1.48353	0.99619	0.08716	11.43005
40	0.69813	0.64279	0.76604	0.83910	86	1.50098	0.99756	0.06976	14.30067
41	0.71558	0.65606	0.75471	0.86929	87	1.51844	0.99863	0.05234	19.08114
42	0.73304	0.66913	0.74314	0.90040	88	1.53589	0.99939	0.03490	28.63625
43	0.75049	0.68200	0.73135	0.93252	89	1.55334	0.99985	0.01745	57.28996
44	0.76794	0.69466	0.71934	0.96569	90	1.57080	1.00000	0.00000	
45	0.78540	0.70711	0.70711	1.00000					

Goodheart-Willcox Publisher

Cosine (cos) is a number that represents the ratio of the length of the side adjacent to the Angle θ divided by the length of the hypotenuse.

$$\text{Cos } \theta = \frac{\text{Adjacent side}}{\text{Hypotenuse}}$$

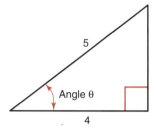

For example, if the length of the adjacent side is 4, and the length of the hypotenuse is 5, then the calculation of the cosine of Angle θ is 4 divided by 5, which equals 0.8. The formula would be written as shown.

$$\text{Cos } \theta = \frac{\text{Adjacent side}}{\text{Hypotenuse}}$$

$$= \frac{4}{5}$$

$$= 0.8$$

By using the trigonometric chart, it can be determined that Angle θ is approximately 37°. By using a scientific calculator, a more accurate degree of 36.87° can be found.

Tangent (tan) is a number that represents the ratio of the length of the side opposite Angle θ divided by the length of the side adjacent to Angle θ.

$$\text{Tan } \theta = \frac{\text{Opposite side}}{\text{Adjacent side}}$$

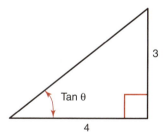

For example, if the length of the opposite side of Angle θ is 3, and the length of the adjacent side is 4, the calculation of the tangent of Angle θ is 3 divided by 4, which equals 0.75. The formula would be written as shown.

$$\text{Tan } \theta = \frac{\text{Opposite side}}{\text{Adjacent side}}$$

$$= \frac{3}{4}$$

$$= 0.75$$

By using the trigonometric chart, it can be determined that Angle θ is approximately 37°. A scientific calculator will find a more accurate reading of 36.87° for Angle θ.

Note that all of the calculations resulted in the same angle. This was done to demonstrate using all of the functions to solve the same triangle. In practice, however, only one of those calculations would be necessary.

In triangles, the sum of the three angles must always equal 180°. In the previous examples, only the right angle (90°) was known. By calculating Angle θ using either the sine, cosine, or tangent, the values of the unknown angle can be determined due to the total of the three angles equaling 180°.

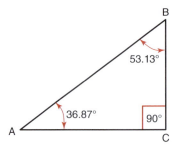

Goodheart-Willcox Publisher

Right angle	90.00°
Angle *A*	36.87°
+ Angle *B*	53.13°
Total	180.00°

In the previous examples, the Angle θ was found by using the lengths of two sides of the triangle. You may not always be provided with the length of two sides; instead, the angle and one of the side lengths may be given. If you know Angle θ and any one of the sides, you can use that information to solve for the rest of the values. The same trigonometric functions are used, but you will use algebraic manipulation to solve for the unknown.

In the triangle pictured, Angle θ is given, which is 40°, and the length of the adjacent side, which is 6. With this information, you can find either the length of the hypotenuse or the opposite side. The cosine function would be used to find the hypotenuse, and the tangent function would be used to find the opposite side. For this example, the tangent function will be used to find the opposite side.

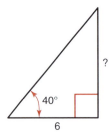

Goodheart-Willcox Publisher

$$\text{Tan } \theta = \frac{\text{Opposite side}}{\text{Adjacent side}}$$

$$= \text{adjacent side} \times \tan \theta$$

$$= 6 \times \tan 40$$

$$= 6 \times 0.83910$$

$$= 5.0346$$

A mnemonic that is often used to help remember the trigonometric functions is **SohCahToa**.

Soh

$$\text{Sin } \theta = \frac{\textbf{O}\text{pposite side}}{\textbf{H}\text{ypotenuse}}$$

Cah

$$\text{Cos } \theta = \frac{\textbf{A}\text{djacent side}}{\textbf{H}\text{ypotenuse}}$$

Toa

$$\text{Tan } \theta = \frac{\textbf{O}\text{pposite side}}{\textbf{A}\text{djacent side}}$$

8.1.2 Application of Sine, Cosine, and Tangent

According to safety standards and ladder manufacturers, a leaning ladder should be placed at a 75° angle. To determine how far the base of the ladder should be placed from the wall, a tringle is drawn, with the hypotenuse being the length of the ladder. For this example, we will use a ladder that is 24′. Since the unknown value is the adjacent side, the cosine function is used.

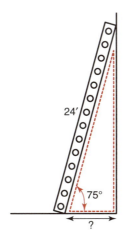

24′

75°

?

Goodheart-Willcox Publisher

$$\text{Cos } \theta = \frac{\text{Adjacent side}}{\text{Hypotenuse}}$$

$$= \text{hypotenuse} \times \cos \theta$$

$$= 24 \times \cos 75$$

$$= 24 \times 0.25882$$

$$= 6.21168′$$

The base of the ladder should be placed approximately 6.21 feet from the wall.

An offset is a common bend made in a conduit. The purpose of an offset is to change the elevation of a conduit or to go over an obstruction. If the scenario is sketched, we can see that a tringle is formed where the conduit changes elevation. The opposite side of the triangle is the height of the obstruction, and the hypotenuse is the distance between bend marks. In this scenario, the elevation change is five inches, and the desired offset angle is 30°. Since the unknown value is the hypotenuse, the sine function will be used.

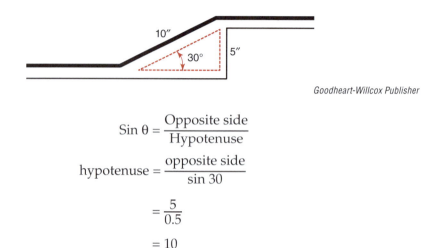

Goodheart-Willcox Publisher

$$\text{Sin } \theta = \frac{\text{Opposite side}}{\text{Hypotenuse}}$$

$$\text{hypotenuse} = \frac{\text{opposite side}}{\sin 30}$$

$$= \frac{5}{0.5}$$

$$= 10$$

The distance between bend marks is 10 inches.

Many benders will have an offset multiplier chart on the bender to save the time of using trig functions to acquire bend marks. The numbers given in the chart are the **cosecant** of the angle, which represents the ratio of the hypotenuse to the opposite side. In the previous example, a 30° bend was used, which had a hypotenuse that was twice the length of the opposite side. Thirty degrees is a common offset angle, and it has a multiplier of two, which provides for easy mental math when determining the distance between bend marks.

Offset Multiplier Chart

Offset Angle in Degrees	Offset Multiplier Obstruction Height
10	6
22.5	2.61
30	2
45	1.41
60	1.15

Goodheart-Willcox Publisher

8.1.3 Pythagorean Theorem

Pythagoras, a Greek mathematician who lived about 2,500 years ago, is credited with discovering the fact that in a right triangle, the square of the hypotenuse is equal to the sum of the squares of the other two sides. This formula is called the **Pythagorean Theorem**. It is written as shown.

$$a^2 + b^2 = c^2$$

where

a = the length of one of the two shorter sides of a right triangle

b = the length of the other shorter side of a right triangle

c = the length of the hypotenuse

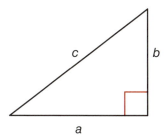

Goodheart-Willcox Publisher

The Pythagorean Theorem is used when performing ac electrical theory calculations involving resistors and inductors or capacitors. On a job site, it is used to ensure that corners are a right angle (90°). By using a triangle that measures in increments of 3, 4, and 5, where 3 and 4 are the short sides and 5 is the hypotenuse, a right angle is proven. Thus, if a measures 3′ and b measures 4′, then c must be 5′.

$$3^2 + 4^2 = 5^2$$

$$9 + 16 = 25$$

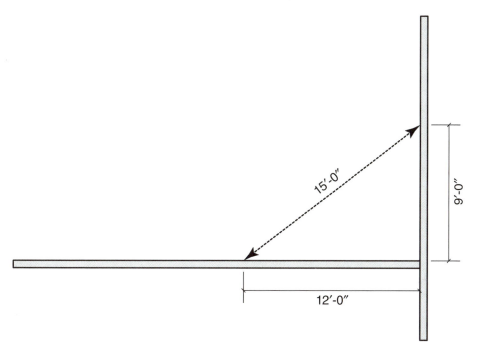

Goodheart-Willcox Publisher

In the diagram shown, if one wall is marked at a position of 9′ from the corner, and the other wall is marked at 12′ from the corner, a measurement between the two marks will equal 15′ if the corner is square (a 90° angle). Any multiples of 3, 4, and 5 can be used in this manner to prove that a corner is a right angle.

$$9^2 + 12^2 = 15^2$$

$$81 + 144 = 225$$

8.1.4 Application of Pythagorean Theory

The impedance of an ac circuit that has a resistor in series with a capacitor or inductor can be solved by using right triangles. The tringle will have the resistance along the bottom, and the inductive or capacitive reactance will be along the side. The hypotenuse of the triangle will be the impedance of the circuit. The Pythagorean Theorem is the fastest way to solve this problem, but trig functions could also be used.

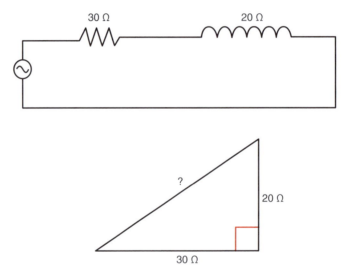

$$c^2 = a^2 + b^2$$

$$c^{\not{2}} = \sqrt{a^2 + b^2}$$

$$= \sqrt{30^2 + 20^2}$$

$$= 36.056$$

The impedance of the circuit is 36.056 ohms.

Review Questions

Side and Angle Exercises

Use the information provided in the triangle shown for Exercises 8-1, 8-2, and 8-3. Find the measurement of the unknown angle or side by using the trigonometric chart found earlier in the chapter or by using a scientific calculator.

Goodheart-Willcox Publisher

Exercise 8-1

Solve the following problems for the unknown angle. Round answers to the nearest tenth of a degree.

1. If Side A measures 10 and Side C measures 15, what is the measurement of Angle 2?

   _____

2. If Side B measures 11.1 and Side C measures 15, what is the measurement of Angle 1?

   _____

3. What is the total measurement of the three angles of this triangle?

   _____

4. If Side A measures 8.5 and Side B measures 21, what is the measurement of Angle 1?

   _____

(Continued)

5. If Side A measures 6.2 and Side B measures 22.75, what is the measurement of Angle 1?

   _____

6. If Side B measures 16 and Side C measures 20, what is the measurement of Angle 1?

   _____

7. If the measurement of Angle 1 is 26.4°, what is the measurement of Angle 2?

   _____

8. If Side B measures 15 and Side C measures 20, what is the measurement of Angle 2?

   _____

9. If Side A measures 32 and Side B measures 50, what is the measurement of Angle 2?

   _____

Exercise 8-2

Solve the following problems for the unknown side. Round answers to the nearest hundredth.

1. If the measurement of Angle 1 is 27° and Side A is 20, what is the length of Side C?

   _____

2. If the measurement of Angle 1 is 55° and Side B is 4, what is the length of Side C?

   _____

Name _____ **Date** _____ **Class** _____

3. If the measurement of Angle 1 is 32° and Side A is 52, what is the length of Side C?

   _____

4. If the measurement of Angle 1 is 10° and Side C is 7, what is the length of Side B?

   _____

5. If the measurement of Angle 1 is 62° and Side A is 12.4, what is the length of Side C?

   _____

6. If the measurement of Angle 1 is 45° and Side B is 8, what is the length of Side C?

   _____

7. If the measurement of Angle 1 is 76° and Side A is 40.6, what is the length of Side C?

   _____

8. If the measurement of Angle 2 is 35° and Side A is 10, what is the length of Side C?

   _____

9. If the measurement of Angle 2 is 61° and Side C is 16.5, what is the length of Side A?

   _____

(Continued)

10. If the measurement of Angle 2 is 17° and Side B is 32.2, what is the length of Side C?

   _____

Exercise 8-3

Solve the following problems for the unknown side. Round answers to the nearest hundredth.

1. If Side A is 12 and Side B is 30, what is the length of Side C?

   _____

2. If Side B is 10 and Side A is 6, what is the length of Side C?

   _____

3. If Side A is 14.6 and Side B is 16, what is the length of Side C?

   _____

4. If Side A is 80 and Side B is 100, what is the length of Side C?

   _____

5. If Side B is 16 and Side C is 20, what is the length of Side A?

   _____

6. If Side C is 43 and Side A is 10, what is the length of Side B?

   _____

7. If Side C is 75 and Side B is 25, what is the length of Side A?

   _____

Name _____   **Date** _____   **Class** _____

8. If Side A is 35.7 and Side B is 26.4, what is the length of Side C?

   _____

9. If Side C is 42 and Side B is 28.5, what is the length of Side A?

   _____

10. If Side B is 100 and Side A is 100, what is the length of Side C?

   _____

Practical Exercise 8-4

Solve the following practical problems. Round answers to the nearest tenth.

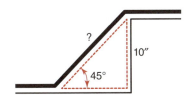

Goodheart-Willcox Publisher

1. Calculate the distance between bend marks (hypotenuse) for a 10″ offset using 45° bends.

   _____ inches

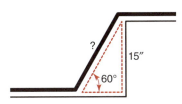

Goodheart-Willcox Publisher

2. Calculate the distance between bend marks (hypotenuse) for a 15″ offset using 60° bends.

   _____ inches

(Continued)

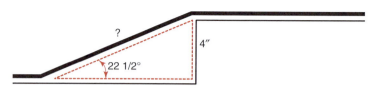

Goodheart-Willcox Publisher

3. Calculate the distance between bend marks (hypotenuse) for a 4″ offset using 22 1/2° bends.

_____ inches

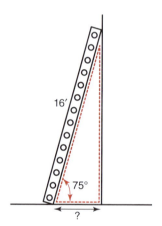

Goodheart-Willcox Publisher

4. Calculate the minimum distance the base of a 16′ ladder must be from a building. The ladder will have a 75° angle with the ground.

_____ feet

Name _____   **Date** _____   **Class** _____

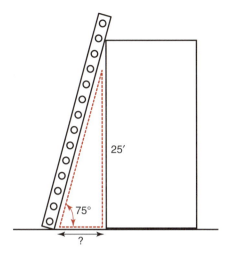

Goodheart-Willcox Publisher

5. Calculate the minimum distance the base of the ladder must be from a building. The ladder will have a 75° angle with the ground.

_____ feet

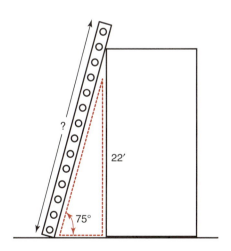

Goodheart-Willcox Publisher

6. Calculate the minimum length of the ladder. The ladder will have a 75° angle with the ground, and it must extend 3′ above the roof.

_____ feet

Practical Exercise 8-5

Solve by applying the given values to the power triangle below. Round answers to the nearest hundredth.

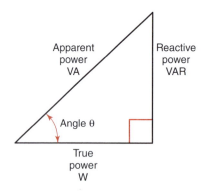

1. Calculate the apparent power of a circuit with a reactive power of 100 VAR and a true power of 80 W.

    _____ VA

2. Calculate the true power of a circuit with an apparent power of 1,000 VA and a reactive power of 725 VAR.

    _____ W

3. Calculate the reactive power of a circuit with an apparent power of 350 VA and a true power of 250 W.

    _____ VAR

4. Calculate Angle θ of a circuit with an apparent power of 4,500 VA and a true power of 4,000 W.

    _____ degrees

CHAPTER 9

Electrical Measurement and Calculation

Objectives

Information in this chapter will enable you to:

- Describe how to measure ac and dc voltage using a multimeter.
- Describe how to measure alternating current and direct current using a multimeter.
- Describe how to measure resistance using a multimeter.
- Calculate voltage, current, and resistance using Ohm's law.
- Calculate power, voltage, and current using Watt's law.
- Describe power factor and the relationship among true power (P), reactive power (Q), and apparent power (S).
- Identify series, parallel, and series-parallel circuits.
- List the characteristics of a series circuit.
- List the characteristics of a parallel circuit.
- Use Ohm's law and Watt's law to calculate voltage, current, resistance, and power in a series, parallel, and series-parallel circuit.

Technical Terms

alternating
 current (ac)
ampere (A)
apparent power
 (S)
capacitive
 reactance
continuity
current (I)
diode check

direct current
 (dc)
electrical circuit
hertz (Hz)
inductive
 reactance
multimeter
ohm (Ω)
Ohm's law
Ohm's law wheel

parallel circuit
power (P)
power factor
reactive power
 (Q)
resistance (R)
series circuit
series-parallel
 circuit
square root

true power (P)
VA (volt-amp)
VAR (volt-amp-
 reactive)
volt (V)
voltage (E)
voltmeter
watt (W)
Watt's law

9.1 Introduction to Electrical Measurement

For electricians, the proper use and understanding of electrical measurement is critical for the safe application of electrical test instruments. Misunderstanding a reading or improper use of an electrical test instrument can result in serious injury or death.

WARNING

This text is not meant to replace or supersede the information provided by the manufacturer of the test instruments used in any application. It is the responsibility of the technician to always ensure the proper instrument is used in a safe manner. Failure to follow the test procedures recommended by the manufacturer and the safety rules of *NFPA 70E* may result in serious injury or death by electrocution.

"Employees working in areas where electrical hazards are present shall be provided with, and shall use, PPE that is designed and constructed for the specific part of the body to be protected and for the work to be performed." *NFPA 70E 130*.

9.2 Electrical Values

There are three major electrical values that are typically measured when working on electrical equipment:

- Voltage
- Current
- Resistance

The electromotive force (EMF) that causes electrons to flow through a conductor is called **voltage (E)**, and it is measured in **volts (V)**. Think of voltage as electrical pressure that causes electrons to move.

The flow of electrons through a conductor is called **current (I)**. The flow of electric current is measured in **amperes (A)**. In the field, this is commonly shortened to *amps*.

The opposition to current flow through a conductor is called **resistance (R)**. The amount of resistance in a circuit or component is measured in **ohms (Ω)**.

9.2.1 Measuring Voltage

Voltage is measured with a **voltmeter**, which is very often just one function of a **multimeter**. Multimeters are instruments that can measure multiple electrical variables. Often, these include voltage, current, and resistance. Multimeters are generally digital, although some analog meters are still in use. A voltmeter may be analog or digital, as well.

Any multimeter used by an electrician should be at least CAT III (600 V) safety rated, and it should be able to measure voltage, current, resistance, continuity, frequency, and capacitance. Additional measurement features may also be available, such as dc millivolts, diode check, and temperature.

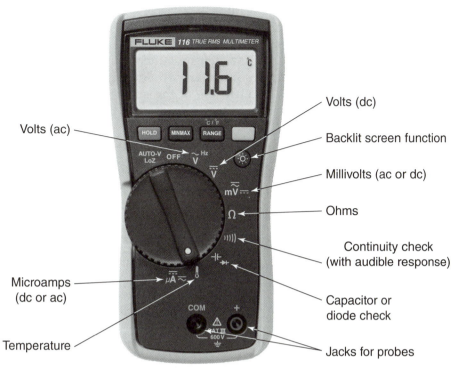

Volts (dc)

Volts (ac)

Backlit screen function

Millivolts (ac or dc)

Ohms

Continuity check
(with audible response)

Microamps
(dc or ac)

Capacitor or
diode check

Temperature

Jacks for probes

Fluke Corporation

The expected voltage of a circuit should be known before testing with a meter. If the voltmeter has multiple voltage ranges, it is good practice to start with the highest voltage range to prevent damage to the meter. Most of the time, electricians will be using the Volts ac setting. Some meters have a millivolt setting that can be used for measuring small voltages.

A voltmeter will typically have settings for ac (alternating current) voltage (Vac) and dc (direct current) voltage (Vdc). Prior to touching the probes to the component to be tested, always check to make sure the meter is set for the proper type of voltage in the circuit.

Note that Vdc is polarity-sensitive. The red probe should be connected to the positive side of the power source, and the black probe should be connected to the negative side of the power source. If these probes are reversed, the measurement will show a negative value. Vac is not polarity-sensitive, and probes can be placed without regard to negative or positive.

After ensuring the proper use of the correctly rated meter, place the electrodes on the contact points of the component to be tested. The meter will display the measured voltage.

9.2.2 Measuring Current

Current is measured in amps, and it is either alternating or direct. **Alternating current (ac)** is an electrical signal that regularly reverses its electron flow. It alternates between positive and negative along a cycle. **Direct current (dc)** is an electrical signal that is steady and maintains a set polarity with electron flow in only one direction. Before measuring current, determine which type of current is used in the circuit to be measured.

Current can be measured with a multimeter, an in-line ammeter, or a clamp-on ammeter. An in-line ammeter requires the electrical circuit to be opened so the ammeter can be connected in line with the load it is measuring. To avoid having to disconnect any wiring, technicians often use a clamp-on ammeter.

9

With a clamp-on ammeter, the reading is taken by placing the meter's clamp around only one of the conductors that feeds the equipment. The meter measures the magnetic field around the conductor. A magnetic field is produced by the current flowing through the conductor. This magnetism is then converted to a reading of amps by the meter. Older clamp-on ammeters would only measure alternating current, but many newer meters are capable of measuring both alternating and direct current.

Clamp-on ammeter

Fluke Corporation

9.2.3 Measuring Resistance

Resistance is measured with a multimeter set on the ohm scale. If the resistance of a single component is to be measured, that component must be removed from the electrical circuit. Voltage cannot be present when measuring resistance, as it could damage the meter. Resistance measurement is commonly used to check for a short circuit, ground fault, and open or a higher than expected resistance.

9.2.4 Additional Measurements

Along with the measurement of voltage, current, and resistance, electricians will commonly encounter the need to check the continuity of a circuit. **Continuity** means there is a complete path for the electrons to flow through a circuit. When a circuit does not have continuity, there is a break in the circuit. This could be caused by a broken connection, an open switch, or a tripped safety device. Most multimeters will have a setting that gives an audible beep if the circuit has continuity.

A **diode check** may also be necessary to determine if a diode is conducting current properly. A diode, like a check valve in a plumbing system, should allow flow only in one direction.

Capacitors are used to start and run electric motors more efficiently and for power factor correction. Motors may have a *start capacitor*, *run capacitor*, or combination *start/run capacitor*. Capacitors are rated in farads, but their capacity is usually low, so the microfarad measurement scale is used.

Frequency is the number of cycles per second in an ac electrical circuit. In the United States, ac current operates at 60 cycles, meaning 120 reversals of direction per second. Frequency is measured in **hertz (Hz)**.

9.3 Electrical Calculations

It may be necessary to calculate certain electrical values, such as the amount of current a piece of equipment will draw, the resistance of a circuit, or the power demand of a particular load. An understanding of the relationship between electromotive force, current, and resistance is imperative, as is knowledge of the difference between series and parallel circuits.

9.3.1 Ohm's Law

Ohm's law, named after nineteenth century German physicist Georg Ohm, describes the relationship between electromotive force, current, and resistance. Ohm determined by measurement that there is a direct relationship among these values in an electrical circuit.

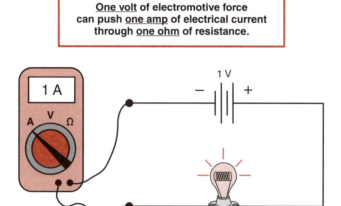

Ohm's Law

One volt of electromotive force can push one amp of electrical current through one ohm of resistance.

Goodheart-Willcox Publisher

Ohm's law can be stated mathematically as shown.

$$E = I \times R$$

where

    E = electromotive force in volts (V)

    I = intensity of the current flow in amps (A)

    R = the resistance to current flow in ohms (Ω)

To find the electromotive force (E, measured in volts), simply multiply the current flow (I, measured in amps) by the resistance (R, measured in ohms).

    To solve for I, the equation can be rewritten as shown.

$$\frac{E}{R} = I$$

To solve for R, the equation would be as shown.

$$\frac{E}{I} = R$$

Equations, such as Ohm's law, can always be changed in this manner by isolating the value to be determined on one side of the equals sign. Therefore, if the formula is written as $E = I \times R$, and the unknown value is I, the formula can be changed by dividing both sides of the equation by R.

9

$$E = I \times R$$

$$\frac{E}{R} = \frac{I \times R}{R}$$

Since there is now an *R* both above and below the dividing line on the right side of the equals sign, they cancel each other.

$$\frac{E}{R} = \frac{I \times \cancel{R}}{\cancel{R}}$$

This leaves the *I* alone on the right side of the equals sign, making the calculation to solve for *I* to be *E* divided by *R*.

$$\frac{E}{R} = I$$

The same process can be used to solve for resistance.

$$E = I \times R$$

The variable *R* can be isolated by dividing both sides of the equation by *I*.

$$\frac{E}{I} = \frac{I \times R}{I}$$

With an *I* both above and below the dividing line on the right side of the equation, they cancel each other.

$$\frac{E}{I} = \frac{\cancel{I} \times R}{\cancel{I}}$$

Therefore, the formula for resistance is *E* divided by *I* equals *R*.

$$\frac{E}{I} = R$$

Many people rely on the **Ohm's law wheel** to help them remember the various formulas.

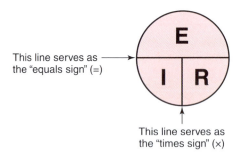

This line serves as the "equals sign" (=)

This line serves as the "times sign" (×)

Goodheart-Willcox Publisher

Thus, when viewed as a wheel, the equation still reads $E = I \times R$. To find any unknown value when two values are known, simply cover up the unknown on the wheel, and the formula is easily revealed. For example, to find *E*, the formula is $I \times R$.

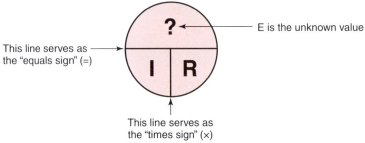

E is the unknown value

This line serves as the "equals sign" (=)

This line serves as the "times sign" (×)

Goodheart-Willcox Publisher

To find *R*, the formula is *E* divided by *I*.

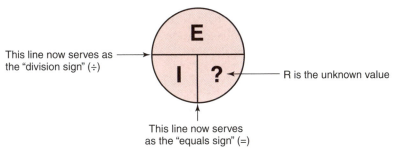

This line now serves as
the "division sign" (÷)

R is the unknown value

This line now serves
as the "equals sign" (=)

Goodheart-Willcox Publisher

To find *I*, the formula is *E* divided by *R*.

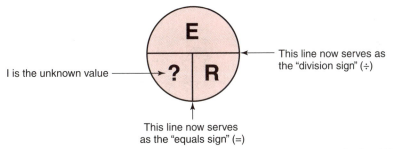

I is the unknown value

This line now serves as
the "division sign" (÷)

This line now serves
as the "equals sign" (=)

Goodheart-Willcox Publisher

For example, to calculate the current flow in a circuit that uses 9 volts of electromotive force to power a load with 36 ohms of resistance, the equation would be as shown.

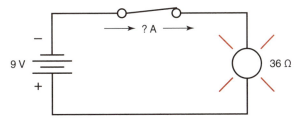

Goodheart-Willcox Publisher

$$I = \frac{E}{R}$$

$$= \frac{9 \text{ V}}{36 \text{ }\Omega}$$

$$= 0.25 \text{ A}$$

Thus, 9 V can push 0.25 amps of current through 36 Ω of resistance.

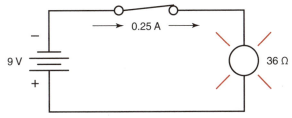

Goodheart-Willcox Publisher

9.3.2 Power Formula

Power (P) can be defined as the rate at which work is being done. Power is measured in **watts (W)**, named in honor of eighteenth century inventor James Watt. Because the watt is a relatively small unit of measurement, it is typically expressed in *kilowatts (kW)*, which is 1,000 watts, or *megawatts (MW)*, which is 1,000,000 watts.

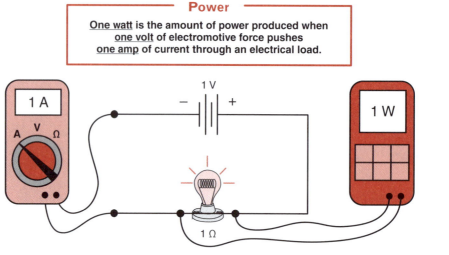

Power

One watt is the amount of power produced when one volt of electromotive force pushes one amp of current through an electrical load.

Goodheart-Willcox Publisher

Just as Ohm's law was developed to describe the relationship among voltage, current, and resistance, **Watt's law** was developed to describe the relationship among power, current, and voltage. It can be stated mathematically as shown.

$$P = I \times E$$

where

P = power in watts (W)

I = intensity of the current flow in amps (A)

E = electromotive force in volts (V)

To find the power consumed by a load (P, measured in watts), simply multiply the current flow (I, measured in amps) by the electromotive force (E, measured in volts). Like Ohm's law, Watt's law can be rewritten to find any one of the three unknown values. To solve for I, the equation can be rewritten as shown.

$$I = \frac{P}{E}$$

To solve for E, the equation would be written as shown.

$$E = \frac{P}{I}$$

Like with Ohm's law, a wheel can be used to help remember the formula for solving for the unknown value in Watt's law.

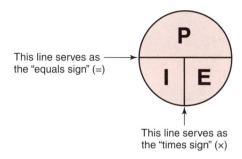

This line serves as the "equals sign" (=)

This line serves as the "times sign" (×)

Goodheart-Willcox Publisher

Thus, when viewed as a wheel, the equation still reads $P = I \times E$. To find any unknown when two values are known, simply cover up the unknown on the wheel, and the formula is easily revealed. For example, to find P, the formula is $I \times E$.

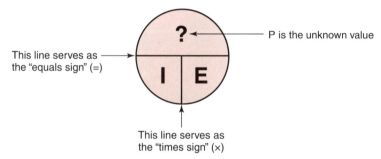

P is the unknown value

This line serves as the "equals sign" (=)

This line serves as the "times sign" (×)

Goodheart-Willcox Publisher

To find E, the formula is P divided by I.

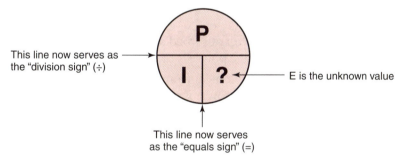

This line now serves as the "division sign" (÷)

E is the unknown value

This line now serves as the "equals sign" (=)

Goodheart-Willcox Publisher

To find I, the formula is P divided by E.

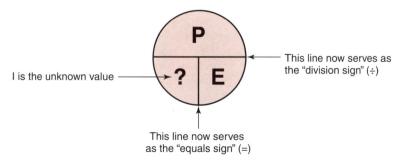

I is the unknown value

This line now serves as the "division sign" (÷)

This line now serves as the "equals sign" (=)

Goodheart-Willcox Publisher

For example, to calculate the power in a circuit that uses 120 volts of electromotive force while measuring 12 amps, the equation would be as shown.

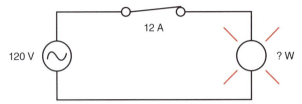

$$P = I \times E$$
$$= 120 \text{ V} \times 12 \text{ A}$$
$$= 1{,}440 \text{ W}$$

Thus, a 120-volt circuit with a current flow of 12 amps has a load that is using 1,440 watts.

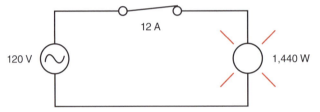

9.3.3 Power Factor

Power in an ac electrical circuit can be thought of in three ways.

- **True power (*P*)** (or *real power*) is the capacity for a circuit to perform work, and it is the amount of power actually used by the load. True power is measured in watts (W).
- **Reactive power (*Q*)** is the power absorbed and returned by the circuit without doing any useful work. Reactive power is measured in **VAR (volt-amp-reactive)**.
- **Apparent power (*S*)** is the total power (combination of true power and reactive power) in the circuit. Apparent power is measured in **VA (volt-amps)**.

The power triangle can be used to demonstrate the relationship of true, apparent, and reactive power. With any two of the three factors known, trigonometry can be used to calculate the unknown value.

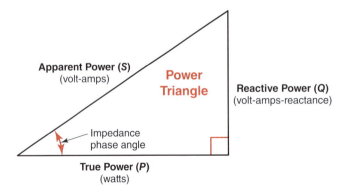

Power factor is the ratio of *true power* (or real power) to *apparent power*. It is used in both single-phase and three-phase power calculations, as it takes into consideration the opposition to current flow. When this opposition is caused by magnetic fields, such as motors, it is called **inductive reactance**. When opposition to current flow is caused by the storage of electrons in an electrostatic field by a capacitor, it is called **capacitive reactance**.

If the load in a circuit is primarily resistive, such as electric heat or incandescent bulbs, then the power factor may be very nearly perfect, or 1.0. Power factor is often referred to as a percentage. A circuit that is nearly perfect and has a 1.0 power factor is describing a 100% power factor. To convert the power factor number to a percentage, it is multiplied by 100, which moves the decimal two places to the right.

If the load is inductive, such as motors and compressors, the power factor may be well below 1.0 or 100%. The lower the power factor, the greater the amount of current that must be carried to handle the circuit's load.

Apparent power will be higher than true power unless the circuit is perfect (power factor = 1), when both numbers will be exactly the same.

The formula for power factor is as shown.

$$\text{Power factor} = \frac{\text{True power (watts)}}{\text{Apparent power (volt-amps)}}$$

In single-phase circuits, the power formula thus becomes:

$$P = I \times E \times \text{power factor}$$

In three-phase circuits, the formula accounts for the additional power available from all the phases; therefore, an additive of 1.73 is used. Note that 1.73 is the **square root** of 3 ($\sqrt{3} = 1.73$).

$$P = I \times E \times 1.73 \times \text{power factor}$$

A three-phase circuit feeds a load that draws 10 amps per phase, has a voltage of 208 volts, and a power factor of 87% (0.87).

$$P = 10 \times 208 \times 1.73 \times 0.87$$

$$= 3,130.61 \text{ watts}$$

9.4 Electrical Circuits

An **electrical circuit** is a path designed to carry current from the electrical source to the load and back. They can be built as series circuits, parallel circuits, or series-parallel circuits (sometimes called *combination circuits*).

- **Series circuits** have just one path for electricity to flow. All the current in a circuit conducts along this single path. Series circuits are used for controls and safety devices, such as switches, relays, overloads, and limits.
- **Parallel circuits** have multiple individual paths to and from the power source for current to flow. The current divides among these paths, and some electrons follow one path, while other electrons follow the other paths. Parallel circuits are used to power electrical loads.
- **Series-parallel circuits** have a series portion and a parallel portion. The series portion includes the controls and safety devices, and the parallel portion provides power to the loads. HVACR equipment most often uses series-parallel circuits.

9.4.1 Series Circuits

Series circuits are primarily used for switches, relays, contactor coils, overloads, and limits because there is a single path for electron flow. Wired in this manner, each of the devices has the same authority to make or break the circuit. There are no devices that can override these devices. In the following diagram, each of the controls is wired in series with each other and in series with the load.

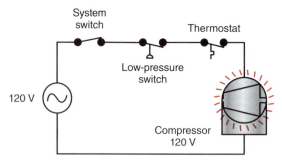

Goodheart-Willcox Publisher

Because all of the switches (system switch, low-pressure switch, and thermostat) are shown closed, power will flow to the compressor. If a single switch opens, such as the thermostat (see the following diagram), power will not flow to the load. A series circuit has only one path for electricity to flow.

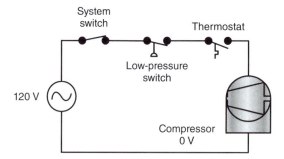

Goodheart-Willcox Publisher

9.4.2 Series Circuit Calculations

In a series circuit, the same amount of current (amperage) flows through the entire circuit. Thus, the same amount of current passes through each component in the circuit.

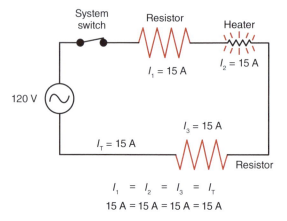

$$I_1 = I_2 = I_3 = I_T$$
$$15\ A = 15\ A = 15\ A = 15\ A$$

Goodheart-Willcox Publisher

In a series circuit, voltages are additive, meaning that the total applied voltage of the circuit is divided up among the loads wired in series. Each electrical load will have a voltage drop that is equivalent to the current passing through the device multiplied by the device's resistance.

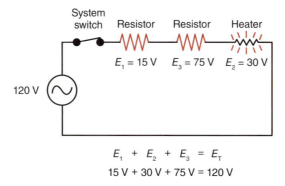

$$E_1 + E_2 + E_3 = E_T$$
$$15\ V + 30\ V + 75\ V = 120\ V$$

Having a shared current and divided voltage are electrical characteristics that are undesirable in most applications. Most electrical loads are designed to operate on a specific voltage that is not intended to be affected by other loads in the circuit.

Like voltage drops, resistances in a series circuit are also additive. Thus, for calculating the total resistance, the resistance of each load is simply added to the others.

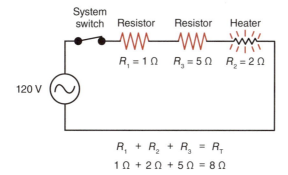

$$R_1 + R_2 + R_3 = R_T$$
$$1\ \Omega + 2\ \Omega + 5\ \Omega = 8\ \Omega$$

9.4.3 Parallel Circuits

In parallel circuits, there is more than one path for electrons to flow. This allows access for multiple loads in the same circuit to the entire supplied voltage in the circuit. As shown in the following diagram, all three loads receive the same voltage, as they are wired in parallel.

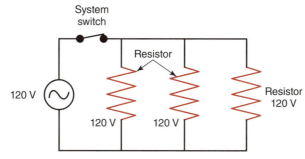

If the system switch is open, as shown in the following diagram, the current flow to all three loads stops. Because the switch is wired in series to all of the loads and it is open, the circuit is broken. Therefore, voltage is not applied to the loads, so no current can flow.

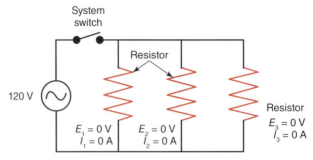

9.4.4 Parallel Circuit Calculations

In a parallel circuit, all of the loads have the same amount of voltage, but the total circuit current is divided among the loads. Therefore, the total current flow (I_T) is the sum of the amperage in each of the branch circuits.

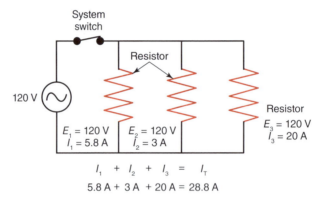

The total resistance (R_T) of the circuit is calculated using Ohm's law. Divide total voltage (E_T) by total current (I_T).

$$R_T = \frac{E_T}{I_T}$$

$$= \frac{120 \text{ V}}{28.8 \text{ A}}$$

$$= 4.16 \text{ }\Omega$$

The resistance of each electrical load can be calculated using Ohm's law. Divide the applied voltage by the current flowing through the electrical load.

$$R_1 = \frac{E_1}{I_1} \qquad\qquad R_2 = \frac{E_2}{I_2} \qquad\qquad R_3 = \frac{E_3}{I_3}$$

$$= \frac{120 \text{ V}}{5.8 \text{ A}} \qquad\qquad = \frac{120 \text{ V}}{3 \text{ A}} \qquad\qquad = \frac{120 \text{ V}}{20 \text{ A}}$$

$$= 20.69 \text{ }\Omega \qquad\qquad = 40 \text{ }\Omega \qquad\qquad = 6 \text{ }\Omega$$

In series circuits, resistances are strictly additive. However, in parallel circuits, this is not the case. Total resistance in a parallel circuit is always less than the resistance of any of the individual branches. This is because current has multiple paths to follow, which increases the overall conductivity of the circuit. When electrons have more ways to move through a circuit (higher conductivity), overall resistance decreases. There are three formulas that can be used to find the total resistance in a parallel circuit.

Equal Value Formula

The equal value formula can only be used when all the resistors connected in parallel have an equal value. The individual value of the resistors is divided by the number of resistors (N) to find the total resistance.

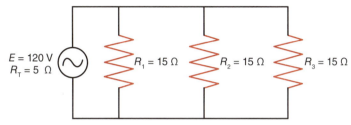

$$R_T = \frac{R}{N}$$

$$= \frac{15}{3}$$

$$= 5 \ \Omega$$

Product over Sum Formula

The product over sum formula can be used when there are two resistors connected in parallel. The product of the two resistors is divided by the sum of the two resistors to find the total resistance.

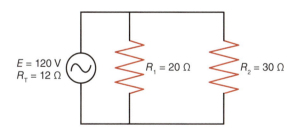

9

$$R_T = \frac{R_1 \times R_2}{R_1 + R_2}$$

$$= \frac{20 \times 30}{20 + 30}$$

$$= \frac{600}{50}$$

$$= 12 \ \Omega$$

Reciprocal Formula

The reciprocal formula can be used in all cases to find the total resistance of resistors connected in parallel. The reciprocal of each resistor is added together. The reciprocal of that sum is the total resistance.

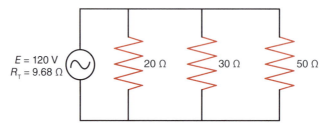

$E = 120$ V
$R_T = 9.68 \ \Omega$
20 Ω   30 Ω   50 Ω

Goodheart-Willcox Publisher

$$\frac{1}{R_T} = \frac{1}{R_1} + \frac{1}{R_2} + \frac{1}{R_3}$$

or

$$R_T = \frac{1}{\dfrac{1}{R_1} + \dfrac{1}{R_2} + \dfrac{1}{R_3}}$$

$$= \frac{1}{\dfrac{1}{20} + \dfrac{1}{30} + \dfrac{1}{50}}$$

$$= 9.68 \ \Omega$$

9.4.5 Series-Parallel (Combination) Circuits

Most electrical systems and equipment have series-parallel circuits. Control devices such as breakers and switches are in series with loads that are connected in parallel. In the following diagram, the system switch is in series with both fans and the compressor. Fan 1 is in parallel with fan 2 and the compressor. Fan 2 and the compressor are also in parallel. Thermostat 1 is in series with fan 1, and thermostat 2 is in series with fan 2 and the compressor. The pressure switch is in series with the compressor.

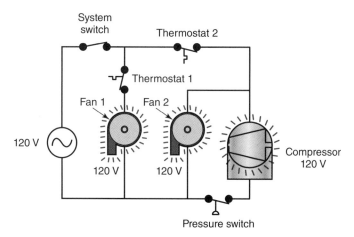

In the previous diagram, all the switches are closed, allowing current to flow through each branch circuit. Thus, the compressor and both fans would be running. In the following diagram, the pressure switch is open, which would not allow current to flow through the compressor. However, both fans would still be running.

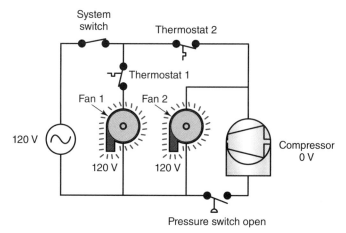

9

9.4.6 Series-Parallel Circuit Calculations

When a series-parallel circuit is analyzed, the series portion follows the rules of a series circuit, and the parallel portion follows the rules of a parallel circuit. The following circuit has a resistor (R_1) that is in series with two resistors (R_2 and R_3) that are connected in parallel.

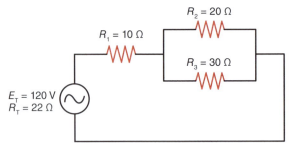

Goodheart-Willcox Publisher

$$R_{2\text{-}3} = \frac{R_2 \times R_3}{R_2 + R_3}$$

$$= \frac{20 \times 30}{20 + 30}$$

$$= 12\ \Omega$$

In the following circuit, the combined value of resistors R_2 and R_3 that are connected in parallel must be found first. The combined value of the two parallel resistors ($R_{2\text{-}3}$) is in series with R_1, so they are added together to find the total resistance (R_T).

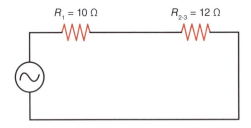

Goodheart-Willcox Publisher

$$R_T = R_1 + R_{2\text{-}3}$$

$$= 10 + 12$$

$$= 22\ \Omega$$

Review Questions

Name _____ Date _____ Class _____

Electrical Measurement Exercise

Exercise 9-1

Using the multimeter pictured, answer the following questions.

1. Is this a bench meter or a clamp-on meter?

   _____

2. What is the CAT rating of the meter (I, II, III, or IV)?

   _____

3. What is the maximum ac voltage rating?

   _____

4. List all the values the multimeter can measure.

   _____

   _____

   _____

   _____

5. What is the maximum amount of current the meter can measure?

   _____

6. What position would the selector switch be placed in to measure frequency?

   _____

7. What position would the selector switch be placed in to measure current when the current value is expected to be around 80 amperes?

   _____

Goodheart-Willcox Publisher

Ohm's Law Exercise

Exercise 9-2

Using Ohm's law, find the missing value in each of the following sets of data. A calculator may be used for this exercise. Round answers to the nearest hundredth.

1. $E = 120$ V, $I = 30$ A, $R =$ _____ Ω

2. $E = 277$ V, $I =$ _____ A, $R = 5.75$ Ω

3. $E =$ _____ V, $I = 15$ A, $R = 7.67$ Ω

4. $E = 208$ V, $I = 40$ A, $R =$ _____ Ω

5. $E =$ _____ V, $I = 40$ A, $R = 5.75$ Ω

6. $E = 240$ V, $I =$ _____ A, $R = 7.67$ Ω

7. $E = 480$ V, $I = 20$ A, $R =$ _____ Ω

8. $E =$ _____ V, $I = 15$ A, $R = 15.33$ Ω

Power Formula Exercises

Exercise 9-3

Using Watt's Law, find the missing value in each of the following sets of data. A calculator may be used for this exercise. Round answers to the nearest hundredth.

1. $E = 120$ V, $I = 30$ A, $P =$ _____ W

2. $E = 120$ V, $I =$ _____ A, $P = 2,160$ W

3. $E =$ _____ V, $I = 15$ A, $P = 1,800$ W

4. $E = 277$ V, $I = 40$ A, $P =$ _____ W

5. $E =$ _____ V, $I = 22$ A, $P = 2,640$ W

6. $E = 480$ V, $I =$ _____ A, $P = 4,400$ W

7. $E = 240$ V, $I = 18$ A, $P =$ _____ W

8. $E =$ _____ V, $I = 30$ A, $P = 6,900$ W

Name _____ **Date** _____ **Class** _____

Practical Exercise 9-4

Find the missing value for the following scenarios. A calculator may be used for this exercise. Round answers to the nearest hundredth.

1. A 1,500-watt baseboard heater is connected to a 240-volt circuit. How much current will the heater draw?

   _____

2. A multimeter is used to measure the voltage and current of a resistive load. The meter indicates that the voltage is 117 volts, and the current draw is 3.2 amperes. How much power is it consuming?

   _____

3. A water heater has a heating element rated at 4,500 watts when connected to 240 volts. If an ohmmeter is to be used to measure the resistance of the heating element, what is the expected resistance value?

   _____

4. A 10kW garage heater is connected to 240 volts. How much current will the heater draw?

   _____

5. A room has twelve 60-watt incandescent lamps used for general illumination that are connected to a 120-volt circuit. How much current do the lamps draw?

   _____

Power Factor Exercises

Exercise 9-5

Calculate the power factor in each of the following situations. Round answers to the nearest hundredth.

1. True power = 1,200 W, Apparent power = 1,300 VA

   _____

2. True power = 1,750 W, Apparent power = 2,170 VA

   _____

3. True power = 4,550 W, Apparent power = 4,600 VA

   _____

Exercise 9-6

Calculate the power in watts for each of the following circuits. Round answers to the nearest tenth.

1. Phase = 1, E = 208 V, I = 15 A, Power factor= 0.97, Power = _____ W

2. Phase = 1, E = 120 V, I = 30 A, Power factor= 0.92, Power = _____ W

3. Phase = 1, E = 240 V, I = 40 A, Power factor= 0.98, Power = _____ W

4. Phase = 1, E = 277 V, I = 60 A, Power factor= 0.82, Power = _____ W

5. Phase = 3, E = 208 V, I = 60 A, Power factor= 0.99, Power = _____ W

6. Phase = 3, E = 240 V, I = 80 A, Power factor= 1.00, Power = _____ W

7. Phase = 3, E = 480 V, I = 100 A, Power factor= 0.76, Power = _____ W

8. Phase = 3, E = 600 V, I = 200 A, Power factor= 0.94, Power = _____ W

Name _____   **Date** _____   **Class** _____

Series Circuit Exercise

Exercise 9-7

In the following series circuit, find the missing values. A calculator may be used for this exercise. Round answers to the nearest hundredth.

1. $E = 120$ V, $I = 10$ A, $R_1 = 2.5\ \Omega$, $R_2 = $ _____ Ω, $R_3 = 5.7\ \Omega$, $R_T = $ _____ Ω,
   $P = $ ____ W

2. $E = 208$ V, $I = $ _____ A, $R_1 = $ _____ Ω, $R_2 = 20.1\ \Omega$, $R_3 = 15.3\ \Omega$, $R_T = 50\ \Omega$,
   $P = $ _____ W

3. $E = 240$ V, $I = $ _____ A, $R_1 = 8.31\ \Omega$, $R_2 = 11.94\ \Omega$, $R_3 = 8.5\ \Omega$, $R_T = $ _____ Ω,
   $P = $ _____ W

4. $E = $ _____ V, $I = 12$ A, $R_1 = 3.9\ \Omega$, $R_2 = $ _____ Ω, $R_3 = 6.1\ \Omega$, $R_T = 18.33\ \Omega$,
   $P = $ _____ W

Parallel Circuit Exercise

Exercise 9-8

In each of the parallel circuits shown below, find the missing values. A calculator may be used for this exercise. Round values containing decimals to the nearest hundredth.

1. $E = 120$ V, Branch 1: $I = 3.0$ A, Branch 2: $I = 2.5$ A, Total current = _____ A,
   Total resistance = _____ Ω, $P = $ _____ W, Power factor = 1.00

2. $E = $ _____ V, Branch 1: $I = 5.0$ A, Branch 2: $I = 3.1$ A, Branch 3: $I = 7.2$ A,
   Total current = _____ A, Total resistance = 7.84 Ω, $P = $ _____ W,
   Power factor = 0.83

3. $E = 240$ V, Branch 1: $I = 7.5$ A, Branch 2: $I = 2.5$ A, Total current = _____ A,
   Total resistance = _____ Ω, $P = $ _____ W, Power factor = 0.91

4. $E = 277$ V, Branch 1: $I = 8.0$ A, Branch 2: $I = 12.5$ A, Branch 3: $I = 3.3$ A,
   Total current = _____ A, Total resistance = _____ Ω, $P = $ _____ W,
   Power factor = 0.74

Series-Parallel Circuit Exercise

Exercise 9-9

Use the following diagram to complete the series-parallel circuit exercise. A calculator may be used for this exercise. Round answers to the nearest hundredth.

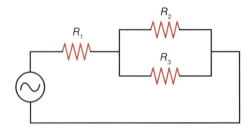

Goodheart-Willcox Publisher

1. $R_1 = 5\ \Omega$, $R_2 = 10\ \Omega$, $R_3 = 2\ \Omega\ 0$, $R_T =$ _____ Ω

2. $R_1 = 3\ \Omega$, $R_2 = 6\ \Omega$, $R_3 = 8\ \Omega$, $R_T =$ _____ Ω

3. $R_1 = 200\ \Omega$, $R_2 = 500\ \Omega$, $R_3 = 300\ \Omega$, $R_T =$ _____ Ω

4. $R_1 = 64\ \Omega$, $R_2 = 100\ \Omega$, $R_3 = 70\ \Omega$, $R_T =$ _____ Ω

CHAPTER 10

Practical Applications

Objectives

Information in this chapter will enable you to:

- Apply concepts and operations involving whole numbers, common fractions, and decimal fractions in electrical applications.
- Apply concepts and operations involving percentages, ratios, and proportions in electrical applications.
- Apply concepts and operations involving algebraic functions, geometric functions, and trigonometric functions in electrical applications.

10.1 Practical Exercises

All of the various mathematical processes explained in this text—from simple arithmetic to complex trigonometry—have practical applications within the electrical industry. The exercises contained in this chapter will replicate real-world situations.

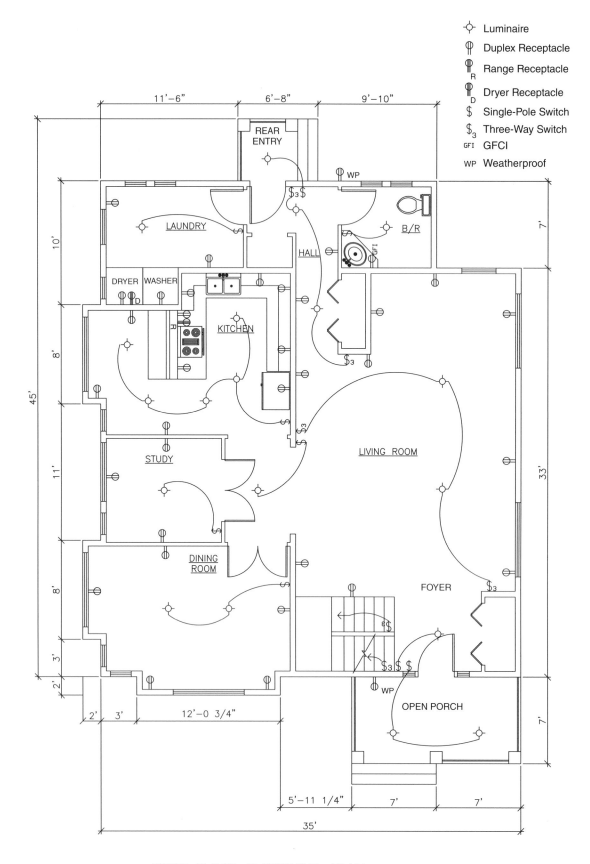

Luminaire
Duplex Receptacle
Range Receptacle
Dryer Receptacle
Single-Pole Switch
Three-Way Switch
GFCI
Weatherproof

REAR ENTRY

LAUNDRY

DRYER | WASHER

KITCHEN

HALL

B/R

STUDY

LIVING ROOM

DINING ROOM

FOYER

OPEN PORCH

FIRST FLOOR ELECTRICAL PLAN
SCALE: 1/4" = 1'-0"

Goodheart-Willcox Publisher

Name _____   **Date** _____   **Class** _____

Practical Exercise 10-1

An electrical contractor is calculating the cost of a job and making a materials list. Use the electrical plan provided to answer the following questions. Only include what is shown on the plan. Do not count switches or lights that would show on another floor.

1. Determine the number of lighting outlets.

   A. Rear Entry: _____

   B. Laundry: _____

   C. Bathroom: _____

   D. Hall: _____

   E. Kitchen: _____

   F. Living Room: _____

   G. Study: _____

   H. Dining Room: _____

   I. Foyer: _____

   J. Open Porch: _____

   K. Total: _____

2. The duplex receptacles in the laundry room, bathroom, kitchen, and dining room are required to be on 20-ampere branch circuits. Determine the number of duplex receptacles that are connected to 20-ampere circuits.

   A. Laundry: _____

   B. Bathroom: _____

   C. Kitchen: _____

   D. Dining Room: _____

   E. Total: _____

3. Determine the number of weatherproof duplex receptacles.

   _____

4. The duplex receptacles in the hall, living room, and study are connected to 15-ampere circuits. Determine the number of duplex receptacles that are connected to 15-ampere circuits.

   A. Hall: _____

   B. Living Room: _____

   C. Study: _____

   D. Total: _____

(Continued)

10

5. Determine the number of single pole switches.

   A. Rear Entry: _____

   B. Laundry: _____

   C. Bathroom: _____

   D. Hall: _____

   E. Kitchen: _____

   F. Living Room: _____

   G. Study: _____

   H. Dining Room: _____

   I. Foyer: _____

   J. Open Porch: _____

   K. Total: _____

6. Determine the number of three-way switches.

   A. Rear Entry: _____

   B. Laundry: _____

   C. Bathroom: _____

   D. Hall: _____

   E. Kitchen: _____

   F. Living Room: _____

   G. Study: _____

   H. Dining Room: _____

   I. Foyer: _____

   J. Open Porch: _____

   K. Total: _____

Name _____   **Date** _____   **Class** _____

Practical Exercise 10-2

The amount of material needed for jobs can be found by determining the number of lighting outlets, receptacles, etc. and multiplying by the average amount of material they use. For example, a duplex receptacle outlet consists of one single-gang box, three straps, three wire connectors, one duplex receptacle, one receptacle plate, and 17 feet of cable. Some receptacle outlets will use more material, and others will use less.

Answer the following questions using the average values for material that has been provided. Use the electrical plan provided with Practical Exercise 10-1 to answer the following questions.

1. The average amount of cable used by a lighting outlet is 17′. The lights in this house are connected to 15-ampere circuits and fed with 14-2 NM cable. How much 14-2 NM cable should be allotted for lighting outlets?

   _____

2. The average amount of cable used by a duplex receptacle outlet is 19′. The receptacle outlets in the laundry room, kitchen, dining room, and bathroom are connected to 20-ampere circuits and fed with 12-2 NM cable. How much 12-2 NM cable should be allotted for the receptacle outlets in these areas?

   _____

3. The average amount of cable used by a duplex receptacle outlet is 19′. The receptacle outlets in the hallway, living room, and study are connected to 15-ampere circuits and fed with 14-2 NM cable. How much 14-2 NM cable should be allotted for the receptacle outlets in these areas?

   _____

4. The average amount of cable used by a weatherproof receptacle outlet is 38′. The weatherproof outlets are connected to 20-ampere circuits and fed with 12-2 NM cable. How much 12-2 NM cable should be allotted for the weatherproof receptacles?

   _____

10

(Continued)

5. Based on the previous calculations, how much 14-2 NM cable will be needed for the first floor?

_____

6. This contractor purchases NM cable in 250-foot rolls. How many rolls of 14-2 NM cable should be provided for the first floor?

_____

7. Based on the previous calculations, how much 12-2 NM cable will be needed for the first floor?

_____

8. The contractor purchases NM cable in 250-foot rolls. How many rolls of 12-2 NM cable should be provided for the first floor?

_____

Name _____   **Date** _____   **Class** _____

Practical Exercise 10-3

A common method for estimating a project is to apply a unit cost to the outlets, devices, etc. The unit cost will include the average amount of materials and labor for each component.

Using the unit cost table provided and the electrical plan from Practical Exercise 10-1, answer the following questions. Only include what is shown on the plan. Do not count switches or lights that would be shown on another floor.

Unit Cost

Unit	Cost
Lighting outlet	$93.00
Single-pole switch	$40.00
Three-way switch	$73.00
Duplex receptacle	$52.00
GFCI receptacle	$95.00
Weatherproof receptacle	$140.00
Dryer receptacle	$420.00
Range receptacle	$440.00

Goodheart-Willcox Publisher

1. Determine the number of GFCI duplex receptacles.

   _____

2. Using the answer from Question 1 and the value in the unit cost table, determine the total cost of the GFCI duplex receptacles.

   _____

3. Using the value in the unit cost table, determine the total cost of the weatherproof duplex receptacles.

   _____

4. Determine the number of duplex receptacles that are *not* GFCI or weatherproof.

   _____

10

(Continued)

5. Using the answer from Question 4 and the value in the unit cost table, determine the total cost of the duplex receptacles.

_____

6. Using the value in the unit cost table, determine the total cost of the lighting outlets.

_____

7. Using the value in the unit cost table, determine the total cost of the single-pole switches.

_____

8. Using the value in the unit cost table, determine the total cost of the three-way switches.

_____

9. Using the previous answers and the unit cost table, determine the total cost of the electrical work for the first floor of the house.

   A. Lighting outlets: _____

   B. Single-pole switches: _____

   C. Three-way switches: _____

   D. Duplex receptacles: _____

   E. GFCI receptacles: _____

   F. Weatherproof receptacles: _____

   G. Dryer receptacles: _____

   H. Range receptacles: _____

   I. Total: _____

Practical Exercise 10-4

A contractor received the following invoice from the supplier for materials for a parking lot light installation.

Electrical Supply Company
2501 3rd Ave N
Fargo, ND 58102
(701) 555-1212

Sold to:	**Shipped to:**
Zach's Electric Inc.	Same
2002 13th Ave S	
Fargo, ND 58102	

Order Date	Date Shipped	Purchase Order #	Account Type	Invoice Date	Terms
5/11	5/27	Z3428	Credit	6/1	1.5% 10 days / Net 30 / 2% added after 30 days

Item #	Quantity	Part #	Mfg.	Desc.	Unit Price	Total
1	12	SSP20	WLO	Light Pole	854.00	10,248.00
2	12	RSX2	Lith	Light Fixture	922.00	11,064.00
3	1	THHN12Bk	SWR	1,000' 12AWG Black	408.00	408.00
4	1	THHN12Rd	SWR	1,000' 12AWG Red	408.00	408.00
5	1	THHN12Gr	SWR	1,000' 12AWG Green	408.00	408.00
6						
7						
8						
9						
10						
Totals	27					$22,536.00

Goodheart-Willcox Publisher

1. The terms of the invoice state there will be a 1.5% discount applied if the bill is paid within 10 days. If the bill is paid on June 9, what amount should be paid?

   _____

2. A finance charge will be added if the bill is not paid on time. What amount will be due if the bill is paid on July 15?

   _____

(Continued)

3. For accounting purposes, the only costs charged to the job will be for the light poles, lights, and 300 feet of each color of wire. What is the cost of these items if paid for within 30 days?

_____

4. The installation of the poles and lights required two electricians. The senior installer worked 46.4 hours at a rate of $37 per hour. The second installer worked 41.6 hours at a rate of $27 per hour. What was the cost of the labor?

_____

5. Travel time is charged at $18 per hour, per vehicle. The installers each drove to the jobsite five days in a row. The senior installer drove 25 minutes each way, and the second installer drove 35 minutes each way. What are the travel charges?

_____

6. What is the total cost of labor and travel charges for this installation?

_____

The following miscellaneous items were used in the installation.

Miscellaneous Items Used

Item	Cost per Unit	Quantity
Fuse holders	$7.54	12
Fuses	$11.30	12
Wire nuts	$0.60	84

Goodheart-Willcox Publisher

7. What is the total cost of the miscellaneous items used?

_____

8. What is the total cost of the labor, travel, and materials for this job?

_____

9. The customer was charged $35,200 for this installation. What was the amount of gross profit on this job?

_____

10. The electrical contractor calculates the overhead on their installation work at 25% of the cost of labor, travel, and materials. With the overhead included, what is the net profit on this job?

_____

Name _____   **Date** _____   **Class** _____

Practical Exercise 10-5

Electrical contractors, estimators, and project managers often perform area and volume calculations when planning jobs. The following are examples of the types of calculations that must be performed. Round answers to the nearest hundredth.

1. A slab on grade rambler has outside dimensions of 35′ × 50′. What is the area of the home?

   _____

2. To calculate the general unit load for a home, the area is multiplied by 3 volt-amps per square foot. What is the general lighting unit load of the home in Question 1?

   _____

3. Calculate the area of the first-floor electrical plan provided in Practical Exercise 10-1. Do not count the open porch or the rear entrance.

   _____

4. Calculate the minimum unit load for the first-floor electrical plan from Practical Exercise 10-1.

   _____

5. Calculate the minimum unit load for the electrical plan from Practical Exercise 10-1 if the home has a basement and second floor that are the same size as the first floor.

   _____

6. A channel had to be cut in the concrete floor of a business to install a floor outlet. The concrete is 4″ thick, and the channel is 12″ wide and 30′ long. What volume of concrete will be needed to repair the floor after the electrical work has been completed?

   _____

10

(Continued)

7. The electrician plans to use bags of concrete to repair the channel from Question 6. Each bag contains 0.45 cubic feet of concrete. How many bags will be needed to fill the trench? Round to the nearest whole number.

_____

8. The electrical contractor is responsible for providing a concrete transformer pad outside of a building. The pad is 20′ long, 8′ wide, and 6″ thick. What is the volume of concrete in cubic yards that will be needed for this installation?

_____

9. A business is installing a solar array. It is a ground-mounted array that is supported by eight posts. Each post is set in a round concrete base that is 2′ in diameter and 6′ deep. What is the volume of concrete in cubic yards that is needed for this installation?

_____

10. A parking lot is to have 26 light poles. Each pole is mounted in a round concrete base that is 20″ in diameter and 8′ deep. What is the volume of concrete in cubic yards that is needed for this installation?

_____

11. A concrete-encased duct bank is to be installed between two industrial buildings. The duct bank is 2′ wide, 3′ deep, and 120′ long. It contains four PVC raceways, each having an outside diameter of 4.5″. What is the volume of concrete in cubic yards that is needed for this installation?

_____

Name _____   **Date** _____   **Class** _____

Practical Exercise 10-6

A retail business has metal halide light fixtures that are starting to fail. The business owner is considering updating the lighting to LED to eliminate frequent repair bills and reduce energy consumption.

The building has 240 metal halide high bay fixtures, each consumes 458 watts. They are on 360 days per year for an average of 17 hours per day.

The cost of new LED fixtures with equivalent output is $195.16 per unit. The new fixtures consume 150 watts each. The cost of labor to install the new fixtures and dispose of the old ones is $180 per unit. The utility company charges 12 cents per kilowatt hour ($0.12 kWh).

1. What is the yearly utility cost to operate the existing lights?

   _____

2. What will be the yearly utility cost to operate replacement LED fixtures?

   _____

3. How much money will be saved per year on utility bills by updating the fixtures?

   _____

4. What is the labor and material cost to replace the fixtures?

   _____

5. Considering yearly energy savings and replacement costs, how long will it take to break even on the fixture upgrade?

   _____

(Continued)

10

6. If 16 of the original fixtures failed per year, at a cost of $370 per fixture to repair, what is the yearly cost to repair the old lights?

_____

7. What is the total yearly cost of the old fixtures? Include the yearly utility cost and the yearly repair costs.

_____

8. Considering energy savings, repair costs, and replacement costs, how long will it take to break even on the fixture upgrade?

_____

Name _____ **Date** _____ **Class** _____

Practical Exercise 10-7

Use trigonometric functions and the Pythagorean Theorem to answer the
following questions. Round answers to the nearest hundredth.

1. An extension ladder is to be used to get on top of a building that is 26′ tall. The
   ladder should be at a 75° angle with the ground. How far will the base of the
   ladder be from the bottom of the wall?

   _____

2. An electrician is setting forms for a transformer pad that measures 8′ × 10′.
   To ensure it is square, a diagonal measurement is taken from two opposite
   corners. What is the diagonal measurement when it is square?

   _____

3. A 45° bend is to be used to make an 11″ offset. What is the distance between
   bend marks?

   _____

4. *Section 358.28(B)* of the *National Electrical Code* limits the amount of bend in an
   electrical metallic tubing (EMT) conduit run to 360°. The EMT conduit already
   has 330° of bend, but it needs an additional 4″ offset. What is the maximum
   angle that can be used without violating the code?

   _____

5. While making the offset from Question 4, what is the distance between bend
   marks?

   _____

(Continued)

10

6. An electrician needs to drill a hole in a roof for a service mast. The center of the hole on the bottom side of the soffit is 22 1/2″ from the front of the fascia board. The roof angle is 33.69°. What is the distance up the roof from the fascia board to the center of the hole?

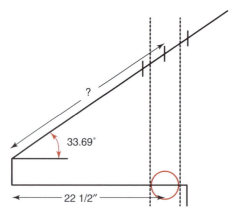

Goodheart-Willcox Publisher

_____

7. Calculate the impedance of a series circuit with 45 ohms of resistance and 30 ohms of inductive reactance.

_____

8. Calculate the inductive reactance of a series circuit with 100 ohms of impedance and 20 ohms of resistance.

_____

9. Calculate the true power of a circuit with 2,000 VA of apparent power and 1,500 VAR of reactive power.

_____

10. Calculate the apparent power of a circuit with 500 W of true power and 400 VAR of reactive power.

_____

Name _____   **Date** _____   **Class** _____

Practical Exercise 10-8

In the following circuits, find the missing values. Round answers to the nearest hundredth.

1. A circuit has four resistors in series. Resistor one is 20 ohms, resistor two is 25 ohms, resistor three is 6 ohms, and resistor four is 42 ohms. What is the total resistance of this circuit?

   _____

2. A series circuit has three equal resistors with a total combined resistance of 30 ohms. What is the value of one of the individual resistors?

   _____

3. A series circuit has a total resistance of 50 ohms. Resistor one is 10 ohms and resistor two is 25 ohms. What is the value of resistor three?

   _____

4. A circuit has three 10-ohm resistors in series. They are connected to a 120-volt circuit. What is the total current value?

   _____

5. Using the circuit from Question 4, how much voltage is dropped in each resistor?

   _____

6. A circuit has four 10-ohm resistors in parallel. What is the total resistance of the circuit?

   _____

10

(Continued)

7. A circuit has three resistors in parallel. Resistor one is 20 ohms, resistor two is 50 ohms, and resistor three is 100 ohms. What is the total resistance of the circuit?

_____

8. A circuit has two resistors in parallel. Resistor one has a current of 12 amperes, and resistor two has a current of 6 amperes. What is the total current of the circuit?

_____

9. A parallel circuit with three resistors has a total current of 13 amperes. Resistor one has a current of 2 amperes, and resistor two has a current of 6 amperes. How much current is resistor three drawing?

_____

10. A parallel circuit with two resistors is connected to a 12-volt source. The value of resistor one is 24 ohms, and resistor two is 6 ohms. What is the current drawn by resistor two?

_____

Name _____ **Date** _____ **Class** _____

Practical Exercise 10-9

Use Ohm's law and Watt's law to answer the following questions. Round answers to the nearest hundredth.

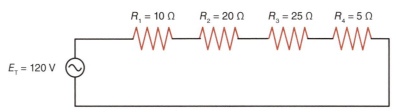

Goodheart-Willcox Publisher

1. Use the series circuit above to complete the calculations.

A. E_1: _____   I. I_T: _____

B. E_2: _____   J. R_T: _____

C. E_3: _____   K. P_1: _____

D. E_4: _____   L. P_2: _____

E. I_1: _____   M. P_3: _____

F. I_2: _____   N. P_4: _____

G. I_3: _____   O. P_T: _____

H. I_4: _____

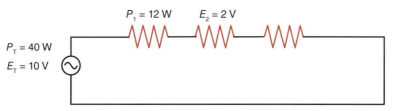

Goodheart-Willcox Publisher

2. Use the series circuit above to complete the calculations.

A. E_1: _____   G. R_1: _____

B. E_3: _____   H. R_2: _____

C. I_1: _____   I. R_3: _____

D. I_2: _____   J. R_T: _____

E. I_3: _____   K. P_2: _____

F. I_T: _____   L. P_3: _____

(Continued)

10

$E_T = 120\ V$ $R_1 = 20\ \Omega$ $R_2 = 30\ \Omega$ $R_3 = 60\ \Omega$

3. Use the parallel circuit above to complete the calculations.

   A. E_1: _____   G. I_T: _____

   B. E_2: _____   H. R_T: _____

   C. E_3: _____   I. P_1: _____

   D. I_1: _____   J. P_2: _____

   E. I_2: _____   K. P_3: _____

   F. I_3: _____   L. P_T: _____

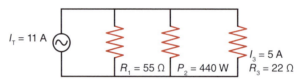

$I_T = 11\ A$ $R_1 = 55\ \Omega$ $P_2 = 440\ W$ $I_3 = 5\ A$ $R_3 = 22\ \Omega$

4. Use the parallel circuit above to complete the calculations.

   A. E_1: _____   G. R_2: _____

   B. E_2: _____   H. R_T: _____

   C. E_3: _____   I. P_1: _____

   D. E_T: _____   J. P_3: _____

   E. I_1: _____   K. P_T: _____

   F. I_2: _____

Name _____ **Date** _____ **Class** _____

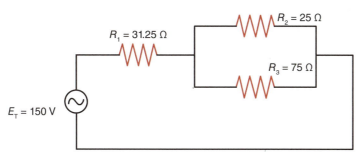

5. Use the series-parallel circuit above to complete the calculations.

A. E_1: _____   G. I_T: _____

B. E_2: _____   H. R_T: _____

C. E_3: _____   I. P_1: _____

D. I_1: _____   J. P_2: _____

E. I_2: _____   K. P_3: _____

F. I_3: _____   L. P_T: _____

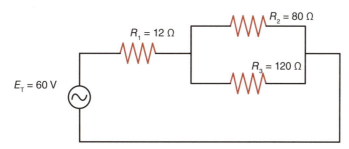

6. Use the series-parallel circuit above to complete the calculations.

A. E_1: _____   G. I_T: _____

B. E_2: _____   H. R_T: _____

C. E_3: _____   I. P_1: _____

D. I_1: _____   J. P_2: _____

E. I_2: _____   K. P_3: _____

F. I_3: _____   L. P_T: _____

Work Space/Notes

Name _____   **Date** _____   **Class** _____

Practical Exercise 10-10

The following are common scenarios encountered in the electrical industry. Use Ohm's law, Watt's law, and other formulas from previous chapters to answer the following questions. Round answers to the nearest hundredth.

1. A 9.6 kVA commercial dishwasher is connected to a 208-volt single-phase branch circuit. How much current will it draw?

   _____

2. A 6,000 W oven is connected to a 240-volt single-phase branch circuit. How much current will it draw?

   _____

3. What is the minimum ampacity of an overcurrent device that feeds a circuit with 12 amperes of continuous load?

   _____

4. What is the maximum amount of continuous load current permitted on a 30-ampere circuit?

   _____

5. A large room has 72 lights, and each draw 0.96 amperes. Light fixtures are likely on for three hours or more, so they are considered a continuous load. What is the minimum number of 20-ampere circuits that are required for this scenario?

   _____

6. A restaurant is installing a booster water heater to preheat the water for the dishwasher. It is a 16-kW water heater, and it connects to three-phase power at 208 volts. How much current will the water heater draw?

   _____

10

(Continued)

7. A plenum heater has several resistive heating elements. An ohmmeter is to be used to measure the resistance of the elements to verify they are operational. The rating of each element is 4,850 watts at 240 volts. What is the expected resistance of the heating elements?

_____

8. A 2,000-watt resistive heater is rated for 240-volt single-phase power. It is connected to a branch circuit that is a considerable distance from the panelboard. It is determined that the conductors feeding the heater have a resistance of 10 ohms. What is the voltage the heater will have while running?

_____

9. The *National Electrical Code* states that a duplex receptacle in a commercial space shall be calculated based on 180 VA. How many duplex receptacles can be installed on a 20-ampere circuit in this scenario?

_____

10. A service calculation for a commercial building has given a value of 523,152 VA. The building is to have a 480-volt three-phase service. What is the minimum ampacity of the service?

_____

Glossary

A

acute angle. An angle that measures more than 0° but less than 90°. (7)

acute triangle. A triangle with all angles less than 90°. (7)

addends. Numbers that are added together. (6)

addition. The process of combining number values to find the sum of those values. (1)

air velocity. A measurement of the distance air moves in a specific length of time. (5)

air volume. The amount of air being moved in a specific length of time. (5)

algebra. A mathematical language used to calculate numerical values in situations where arithmetic alone will not suffice. (6)

alternating current (ac). An electrical signal that regularly reverses its electron flow. (9)

altitude. A length along a perpendicular line from a triangle's base to the opposite angle. (7)

ampere (A). A unit used to measure the flow of electric current. (9)

angle. A figure formed by the intersection of two straight lines emanating from a common point. (7)

angle theta (Angle θ). The angle used within trigonometric functions. (8)

apparent power (S). The total power in an electrical circuit, which is a combination of true power and reactive power. Measured in VA (volt-amps). (9)

arc. A portion of a circle's circumference. (7)

associative property. A mathematical property that allows numbers to be added and multiplied regardless of how they are grouped in an equation. (6)

B

base. Any one side of a triangle. (7)

borrowing. In subtraction, the process of subtracting 1 from a place value and adding 10 to the next-lower place value that needs the 10 in order to be greater than the value of the digit being subtracted from it. (1)

British thermal unit (Btu). A US Customary unit of heat representing the amount of heat energy required to change the temperature of 1 lb of pure water by 1°F. (5)

Btu per hour (Btu/hr). A US Customary unit of heat representing the number of Btus moved in an hour. (5)

C

calorie (cal). An SI unit of heat representing the amount of heat energy required to change the temperature of one gram of water by 1°C. (5)

capacitive reactance. The opposition to current flow caused by the storage of electrons. (9)

capacity. The amount of material a container can hold. (5)

carried over. In addition, a term describing a value transferred from a digit's column to the next-higher digit when such a value is available. (1)

Celsius scale. An SI scale of temperature that uses 0°C as the freezing point of water and 100°C as the boiling point. (5)

chord. A line that touches two points along the circumference of a circle without going through the center. (7)

circle. A closed curve with all points on the curve equidistant from the center. (7)

Note: The number in parentheses following each definition indicates the chapter in which the term can be found.

circumference. The distance around the outside of a circle. (7)

common denominator. A denominator number that is the same for two or more fractions. (2)

commutative property. A mathematical property that allows for the operations of multiplication and addition to be completed in any order. (6)

continuity. Having a complete path for electrons to flow through in a circuit. (9)

cosecant. The ratio of the hypotenuse to the opposite side in a right triangle. (8)

cosine (cos). In a right triangle, a number that represents the ratio of the length of the side adjacent to the angle divided by the length of the hypotenuse. (8)

cube. A value multiplied by itself twice. (1)

cubic feet (ft³). A measure of volume using units of feet that combines length, width, and height. (7)

cubic feet per minute (cfm). A US Customary unit of air volume representing the number of cubic feet of air moved in one minute. (5)

cubic meter per minute (m³/min). An SI unit of air volume representing the number of cubic meters of air moved in one minute. (5)

cubic unit. A unit used to describe volume, such as cubic feet or cubic meters. (7)

current (I). The flow of electrons. (9)

D

decimal. A fractional portion (less than one) of a number expressed using place values to the right of a decimal point. (3)

decimal equivalent. The decimal form of a fraction. (3)

decimal point. A mathematical symbol resembling a period that denotes the place between the whole number and the decimal number. For example: 1.25, where the (.) divides the whole number 1 from the decimal .25 (fraction of 1/4). A decimal point designates where place values change from positive to negative powers of 10. (3)

degree. A unit used in temperature measurement in the common temperature scales. (5)

denominator. The number located below the line in a fraction. (2)

diameter. The distance across a circle going through its center. (7)

difference. The answer to a subtraction equation. (1)

digit. Any of the ten number symbols (0, 1, 2, 3, 4, 5, 6, 7, 8, and 9) in the Arabic numbering system. (1)

diode check. A function available on some multimeters that can determine whether a diode is functioning properly. (9)

direct current (dc). An electrical signal that is steady and maintains a set polarity with electron flow in only one direction. (9)

distributive property. A mathematical property where multiplication of a sum can be accomplished by multiplying each of the addends and then adding the products together. (6)

dividend. The number to be divided in a division equation. (1)

division. The process of separating numbers into groups of smaller numbers. Often thought of as the opposite of multiplication. (1)

division sign (÷). A mathematical symbol indicating to perform division. (1)

divisor. The number of times a number (the dividend) is to be divided in a division equation. (1)

E

electrical circuit. A path designed to carry current from the electrical source to the load and back. (9)

equals sign (=). A mathematical symbol used to denote the answer to a calculation. (1)

equation. A comparison between two or more formulas or values. (6)

equilateral triangle. A triangle with all three sides equal in length and all angles equal (60° each). (7)

exponent. A number or symbol denoting the power to which another number is to be raised. (1)

exterior angle. An angle between the extension of a side of a shape and its adjacent side. (7)

extremes. In a ratio, the two numbers farthest from the equals sign. When expressed in the form of fractions, they are the lower-value numerator and the higher-value denominator. (4)

F

factor (fractions). A whole number that divides evenly into another whole number, especially when used to determine the lowest common denominator (LCD) of two or more fractions. (2)

factor (multiplication). In a multiplication equation, the number being multiplied. (1)

Fahrenheit scale. A US Customary scale of temperature that uses 32°F as the freezing point of water and 212°F as the boiling point. (5)

feet per minute (fpm). A US Customary unit of air velocity, representing the number of feet that air moves in one minute. (5)

foot (ft). A US Customary unit of linear measurement that is equivalent to 12 inches or 1/3 of a yard. (5)

formula. An expression of a mathematical procedure, establishing a method of completing the calculation. (6)

fraction. A portion of a whole number. (2)

G

geometry. A branch of mathematics that deals with the measurement, properties, and relationships of points, lines, angles, surfaces, and solids. (7)

gram (g). An SI unit of weight that is equal to 1/1,000 of a kilogram. (5)

greatest common factor. The greatest possible factor of two or more whole numbers. (2)

H

heat. A measurement of the energy content of a material. (5)

hertz (Hz). A unit used to measure frequency. (9)

I

improper fraction. A fraction in which the numerator is a larger number than the denominator. (2)

inch (in). The base unit of linear measurement in the US Customary system that is equivalent to 1/12 of a foot or 1/36 of a yard. (5)

inches of water column (in. WC). A US Customary unit of pressure used for very small pressure levels and for vacuum and determined by the difference in the heights of two columns of water. (5)

inductive reactance. The opposition to current flow caused by magnetic fields. (9)

interior angle. An angle on the inside of a shape. (7)

isosceles triangle. A triangle with two equal sides. (7)

J

joule (J). A small SI unit of heat. To change the temperature of one kilogram of water by 1°C, 4,187 joules are required. (5)

K

Kelvin scale. An SI scale of temperature that uses absolute zero as its baseline, and each degree has the same value as a Celsius degree. Also known as *Celsius absolute*. (5)

kilogram (kg). A standard unit of weight that serves as the base unit in the SI system. It is equal to 1,000 grams. (5)

kilojoule (kJ). An SI unit of heat. To change the temperature of one kilogram of water by 1°C, 4.187 kilojoules are required. (5)

kilometer (km). An SI unit of linear measurement that is equal to 1,000 meters. (5)

kilopascal (kPa). An SI unit of pressure commonly used in the HVACR field that is equal to 1,000 pascals. (5)

L

line. In geometry, a narrow strip or border that divides or connects areas or objects. (7)

linear measurement. A measurement of the length, width, or depth of an object or the distance between two points. (5)

lowest common denominator (LCD). When comparing two or more fractions, it is the lowest possible denominator to which each fraction can be raised. (2)

lowest terms. The least values to which two or more fractions can be reduced when used in a calculation. (2)

M

mass. The amount of something's matter. (5)

mean. The average value of a group of numbers that is calculated by adding together all the values in the group and then dividing the sum by the number of values in the group. (6)

means. In a ratio, the two numbers closest to the equals sign. When expressed in the form of fractions, they are the higher-value numerator and the lower-value denominator. (4)

median. The midpoint or the middle of the group of values, with half of the values being lower than the median and half of the values being higher than the median. (6)

meter (m). The base unit of linear measurement in the SI system. (5)

meter per minute (m/min). An SI unit of air velocity, representing the number of meters that air moves in one minute. (5)

micron (μm). An SI unit of pressure used for vacuum measurement that is equal to 1/1,000 of a millimeter or 1/1,000,000 of a meter. Also called a *micrometer*. (5)

millimeter (mm). An SI unit of linear measurement that is equal to 1/1,000 of a meter. (5)

minus sign (–). A mathematical symbol indicating to perform subtraction. (1)

mixed number. A value consisting of a whole number and a fraction. (2)

mode. The value in a series of values that is seen most often. (6)

multimeter. An instrument that can measure multiple electrical variables, primarily voltage, resistance, and current. (9)

multiplication. A process of combining number values to get a total by combining factors. (1)

multiplication sign (×). A mathematical symbol indicating to perform multiplication. (1)

N

negative. A term describing numbers with a value less than zero. (1)

numerator. The number located above the line in a fraction. (2)

O

order of operation. The sequence in which mathematical operations should be performed in an equation: parentheses, exponents, multiplication/division, and addition/subtraction (PEMDAS). (1)

obtuse angle. An angle that measures more than 90° but less than 180°. (7)

obtuse triangle. A triangle with one angle that is more than 90°. (7)

ohm (Ω). A unit of electrical resistance. (9)

Ohm's law. A principle that describes the relationship of electromotive force, current, and resistance: $E = I \times R$ (voltage = current × resistance). (9)

Ohm's law wheel. A circular table using Ohm's law to depict the relationship of the variables of voltage, current, and resistance and shows how to calculate unknown values based on known values. (9)

ounce (oz). A US Customary unit of weight that is equal to 1/16 of a pound. (5)

P

parallel circuit. An electrical circuit that has multiple individual paths to and from the power source for current to flow. (9)

parallelogram. A four-sided figure with opposite sides that are parallel. (7)

parentheses (). Structures used in formulas and equations to set apart a certain set of numbers or to indicate which portion of the calculation shall be performed first. (6)

pascal (Pa). An SI unit of pressure that is equal to the amount of force of one newton pushing on one square meter of area. (5)

percentage. A portion of a whole that is expressed by using a percent sign, such as 25% or 50%. (4)

perimeter. The distance around the outside of a shape. (7)

perpendicular. A condition in which two lines form a right angle (90° angle). (7)

pi (π). A mathematical constant (often shortened to 3.14) that describes the relationship between the circumference and the diameter of a circle that never changes. (7)

place value. The location of a digit within a whole number, which determines the value assigned to that digit. (1)

plus sign (+). A mathematical symbol indicating to perform addition. (1)

point. A definite position on a line. (7)

polygon. A two-dimensional closed plane with any number of straight sides. (7)

positive. A term describing numbers with a value greater than zero. (1)

pound (lb). A standard unit of weight that serves as the base unit in the US Customary system. (5)

pounds per square inch (psi). A US Customary unit of pressure, representing the number of pounds of force applied to an area of one square inch. (5)

power (*P*). The rate at which work is being done. (9)

power factor. The ratio of true power (real power) to apparent power. (9)

powers of ten. A method of expressing values that uses exponents to denote to which power of ten a number should be multiplied. (1)

pressure. Force per unit of area and represented by the formula: pressure = force ÷ area. (5)

prime number. A number that has only two factors (1 and itself). (2)

product. The result of a multiplication equation. (1)

proper fraction. A fraction in which the numerator is a smaller number than the denominator. (2)

proportion. A comparison of two ratios, usually separated by an equals sign. (4)

Pythagorean Theorem. A formula concerning right triangles stating that the square of the hypotenuse is equal to the sum of the squares of the other two sides ($a^2 + b^2 = c^2$, where a and b are the lengths of the two shorter sides of a right triangle and c is the length of the hypotenuse). (8)

Q

quotient. The result of a division equation. (1)

R

radius. The distance from the center of a circle to any point along its edge. (7)

Rankine scale. A US Customary scale of temperature that uses absolute zero as its baseline, and each degree has the same value as a Fahrenheit degree. Also known as *Fahrenheit absolute*. (5)

ratio. An expression of a comparison of two or more values that are separated by a colon. (4)

reactive power (*Q*). The power absorbed and returned by a circuit without doing any useful work. Measured in VAR (volt-amp-reactive). (9)

rectangle. A polygon with opposite sides equal in length and adjacent sides unequal in length. (7)

reducing. Lowering a fraction to its lowest possible numerator and denominator. (2)

resistance (*R*). The opposition to the flow of electric current. (9)

right angle. A 90° angle formed by two intersecting lines. (7)

right triangle. A triangle with one angle of 90° and two angles of 45°. (7)

rounding. The process of increasing or decreasing the value of a number to the next digit. (1)

S

scalene triangle. A triangle with no equal sides and no equal angles. (7)

scientific notation. A form of simplified writing of numbers that are too large or too small to be conveniently written in a standard format. (1)

series circuit. An electrical circuit that has just one path for electricity to flow. (9)

series-parallel circuit. An electrical circuit that has a mixture of series and parallel pathways. (9)

SI system. The International System of Units (SI), which is based on standard units for length, weight, and volume with prefixes in powers of ten. Also called the *metric system*. (5)

sine (sin). In a right triangle, a number that represents the ratio of the length of the side opposite any angle divided by the length of the longest side (hypotenuse). (8)

solid. In geometry, a three-dimensional (3D) object, such as a cube, cylinder, pyramid, or sphere. (7)

square (mathematical function). A value multiplied by itself. (1)

square (shape). A polygon that has all four sides equal in length. (7)

square feet (ft²). A measure of area (length × width) in units of feet. (7)

square inch (in²). A measure of area (length × width) that is in units of inches. (5)

square meter (m²). A measure of area (length × width) that is in units of meters. (5)

square root. A divisor (r) of a number (x) that equals the number (x) when the divisor (r) is squared: $r^2 = x$. Therefore, $\sqrt{x} = r$. The square root function is represented by $\sqrt{}$. (9)

square unit. A unit used to describe area, such as square feet or square meters. (7)

straight angle. An angle that is exactly 180°. (7)

subtraction. The process of removing, or taking away, number values from one another to find the difference between those values. (1)

sum. The answer to an addition equation. (1)

superscript. The writing of characters in a smaller size and raised above the bottom line. Commonly used for exponents as seen in squares and cubes: 5^2 and 6^3. (1)

surface. A two-dimensional plane. (7)

T

tangent (in geometry). A straight line intersecting with the circumference of a circle but not entering the circle. (7)

tangent (tan) (in trigonometry). A number that represents the ratio of the length of the side opposite Angle θ divided by the length of the side adjacent to Angle θ. (8)

temperature. The measurement of the intensity of the heat energy in a material. (5)

therm. A US Customary unit of heat equal to 100,000 Btu. (5)

ton (t). A US Customary unit of weight that is equal to 2,000 pounds. (5)

ton of refrigeration. A US Customary unit of heat used as a measure of heat transfer that is equal to the melting of one ton (2,000 lb) of ice at 32°F in 24 hours, which is equal to 288,000 Btu. This can also be reduced to a ton per hour for rating systems, which is equal to 12,000 Btu per hour. (5)

trapezoid. A four-sided polygon, with two parallel sides, and adjacent sides that may or may not be parallel to each other. (7)

triangle. A polygon with three sides. (7)

trigonometry (trig). A branch of mathematics dealing with the angles and sides of triangles and their relationships. (8)

true power (*P*). The capacity for an electrical circuit to perform work and the amount of power actually used by the electrical load. Also called *real power*. Measured in watts. (9)

U

unknown. An unknown value or quantity that is designated by a letter (such as *X*) in an equation or formula. Also called a *variable*. (6)

US Customary system. A standard system of centuries-old measurement units, such as the inch, gallon, and pound. Also called the *inch-pound (IP) system*. (5)

V

VA (volt-amp). A unit representing apparent power (combination of true power and reactive power in an electrical circuit). (9)

VAR (volt-amp-reactive). The unit in which reactive power (*Q*) is measured. (9)

variable. An unknown value or quantity that is designated by a letter (such as *X*) in an equation or formula. Also called an *unknown*. (6)

vertex. A corner of a triangle. (7)

volt (V). A unit used to measure electromotive force (EMF) or voltage (potential electrical difference). (9)

voltage (*E*). The electromotive force (EMF) that causes electrons to flow. (9)

voltmeter. A meter used to measure voltage. (9)

volume. The amount of space that is occupied by a three-dimensional object. (7)

W

watt (W). A unit of power. (9)

Watt's law. A principle that describes the relationship of electrical power, current, and electromotive force: $P = I \times E$ (power = current × voltage). (9)

weight. The value of the gravitational force exerted on a particular piece of matter. (5)

Y

yard (yd). A US Customary unit of linear measurement that is equivalent to 36 inches or 3 feet. (5)

Index

Answers to Odd-Numbered Questions

CHAPTER 1
Whole Numbers

Whole Number Exercises

Exercise 1-1
1. 8
3. 1
5. Hundreds
7. Tens

Exercise 1-2
1. 49,999
3. 799,999

Practical Exercise 1-3
1. 1,750,500

Practical Exercise 1-4
1. 99,999

Comparing Exercise

Exercise 1-5
1. <
3. >
5. <
7. >

Rounding Exercise

Practical Exercise 1-6
1. 900

Addition Exercises

Exercise 1-7
1. 37
3. 125
5. 91
7. 122
9. 488
11. 563
13. 759
15. 24,895
17. 100,981
19. 11,654

Practical Exercise 1-8
1. 197

Practical Exercise 1-9
1. 688

Subtraction Exercises

Exercise 1-10
1. 11
3. 1,783
5. 87,879
7. 889

Practical Exercise 1-11
1. $1,166

Practical Exercise 1-12
1. 2,390 feet

Combined Exercises

Exercise 1-13

1. 108
3. 19,008
5. 59,882
7. 1,672,957,133
9. 324
11. 100,000

Practical Exercise 1-14

1. $1,086 [(3 × 92) + (6 × 135) = 1,086]

Practical Exercise 1-15

1. Regular vehicles daily: 3,400 miles
3. Regular vehicles total: 68,000 miles
5. Emergency Sunday daily: 960 miles
7. Total: 78,560 miles

Division Exercises

Exercise 1-16

1. 5
3. 4 R 4
5. 158 R 6
7. 1,571
9. 4 R 4,068

Practical Exercise 1-17

1. 3 hours

Negative Number Exercise

Practical Exercise 1-18

1. 134°F

Combined Operations Exercises

Practical Exercise 1-19

1. 1,200
3. $110.63 (Rounded up from $110.625)

Practical Exercise 1-20

1. $726.80 [4 filters for each AHU × $7.90 = $31.60 (for each AHU) × 23 AHUs]
3. $17.28 per day [$726.80 (4 filters for each AHU × $7.90 = $31.60 for each AHU × 23 AHUs) + $828 of labor (1/2 hour × 23 AHUs = 11.5 hours × $72/hour) = $1,554.80 ÷ 90 days = $17.28 per day (rounded from $17.2755)]

CHAPTER 2
Fractions

Fractions Exercises

Exercise 2-1

1. 3/4
3. 3/4
5. 3/2 or 1 1/2
7. 3/5
9. 1/3
11. 1/3

Exercise 2-2

1. 3/4 (72/96)
3. 5/16 (30/96)
5. 1/16 (6/96)
7. 1/2 (48/96)
9. 9/16 (54/96)
11. 31/32 (93/96)
13. 11/16 (66/96)
15. 1/1 (96/96)
17. 15/16 (90/96)

Exercise 2-3

1. 1/32: 1/32 (3/96)
3. 5/16: 5/16 (30/96)
5. 15/32: 15/32 (45/96)
7. 34/64: 17/32 (51/96)
9. 5/8: 5/8 (60/96)
11. 8/12: 2/3 (64/96)
13. 9/12: 3/4 (72/96)
15. 7/8: 7/8 (84/96)
17. 31/32: 31/32 (93/96)

Practical Exercise 2-4

1. No, it is not long enough. To compare the two numbers, multiply 5/8 by 2/2 to get 10/16. Since 10/16 is less than 13/16, the piece will not be long enough.

Addition and Subtraction Exercise

Exercise 2-5

1. 1 1/4
3. 3/10
5. 1/16
7. 11 1/4
9. 17 1/16
11. 10 11/16

Multiplication Exercise
Exercise 2-6
1. 7/16
3. 5/18
5. 92 3/16
7. 25/64
9. 11 3/8
11. 1 9/16

Division Exercises
Exercise 2-7
1. 1 (2/2)
3. 2 2/9
5. 1 3/4
7. 21/104
9. 11/16
11. 1 1/2

Practical Exercise 2-8
1. 25

Practical Exercise 2-9
1. 645 3/4 inches [65 1/2 + (11 × 52 3/4) = 645 3/4]

CHAPTER 3
Decimals

Fraction to Decimal Conversion Exercises
Exercise 3-1
1. 0.300
3. 0.778
5. 0.833
7. 0.818
9. 0.955
11. 0.923

Exercise 3-2
1. 5/1,000
3. 3/10
5. 7/9
7. 5/6
9. 11/12
11. 47/50

Decimal to Fraction Conversion Exercises
Exercise 3-3
1. 17/50
3. 7/10
5. 37/500
7. 181/10,000
9. 19/100
11. 17/5,000

Practical Exercise 3-4
1. A. Television: $1.53
   B. Freezer: $12.78
   C. Fridge: $7.10
   D. Range: $9.23
   E. Lights: $3.12
   F. Electric Water Heater: $59.64
   G. Garage Door Opener: $0.80
   H. Dishwasher: $3.55
   I. Washing Machine: $3.76
   J. Clothes Dryer: $6.39
   K. Computer: $0.85
   L. Vacuum: $0.31
   M. Coffee Maker: $1.02
   N. Microwave Oven: $1.08

Decimal Exercises
Practical Exercise 3-5
1. 15.36 A
3. 7.36 A
5. 8 A

Practical Exercise 3-6
1. 16 A [20 × 0.80]

Practical Exercise 3-7
1. 7.8°F (4.38°C)

Practical Exercise 3-8
1. 42.7 hours (42 hours and 42 minutes)
3. $25.36

Practical Exercise 3-9
1. 6,250 watt/hour
3. Zone 1: 6,800 watt/hour, Zone 2: 6,975 watt/hour, Zone 3: 18,750 watt/hour, Zone 4: 36,750 watt/hour, Zone 5: 9,050 watt/hour
5. $15.67

Practical Exercise 3-10
1. 2.6 hours [13 hours ÷ 5 service calls]

Practical Exercise 3-11
1. 330 minutes [480 minutes − 150 minutes of travel]
3. 110 minutes [330 minutes ÷ 3 service calls]

CHAPTER 4
Percentages, Ratios, and Proportions

Conversion Exercise
Exercise 4-1
1. 0.13
3. 0.93
5. 1.0
7. 0.09
9. 0.012
11. 1.101

Percentage Exercises
Exercise 4-2
1. 78%
3. 10%
5. 100%
7. 8.9%
9. 0.32%
11. 3,000%

Practical Exercise 4-3
1. 23%
3. 200 amperes
5. 75.6 amperes

Ratio and Proportion Exercises
Exercise 4-4
1. 10
3. 2
5. 16
7. 28
9. 6
11. 12

Practical Exercise 4-5
1. 24 volts
3. 0.2 amperes

CHAPTER 5
Linear Measurements and Conversions

Linear Measurement Conversion Exercises
Exercise 5-1
1. 144″
3. 252″
5. 136″
7. 163″
9. 268″

Exercise 5-2
1. 2′
3. 15′
5. 7′-5″
7. 5′-3″
9. 11′-2″

Adding and Subtracting Linear Measurements Exercises
Exercise 5-3
1. 4′-12″ or 5′
3. 19′-7″
5. 23′-5″
7. 15′-2″
9. 9′-8″

Exercise 5-4
1. 4′-6″
3. 10′-1″
5. 1′-3″
7. 17′-11″
9. 22′-6″

Multiplying and Dividing Linear Measurements Exercises

Exercise 5-5

1. 5′
3. 36′-8″
5. 44′-4″
7. 21′-3″
9. 6′-1″

Exercise 5-6

1. 2′-3″
3. 1′-11″
5. 8″
7. 2″
9. 2′-6 1/2″

Conversion Exercise

Exercise 5-7

1. 0.3048; 8.2296
3. 3.281; 360.91
5. 0.4732; 5.4418
7. 0.028349; 13.409077
9. 2.2046; 274.03178
11. $(°F-32) \times 5/9$; 33.6°C
13. 34.5; 32,775
15. 0.001; 1,200
17. 252.16; 86,995.2
19. 2.036; −33.594
21. 0.0000102; 0.01275
23. 0.06895; 31.8549
25. 35.31; 8,297.85

Unknown Value Exercise

Exercise 6-2

1. X = 147
3. Z = 492
5. Y = 34
7. X = 176
9. Z = 90.5

Rearranging Formulas Exercises

Exercise 6-3

1. $I = \dfrac{P}{E}$
3. $E = R \times I$
5. $C = \dfrac{1}{(2\pi F \times X_C)}$
7. $Z = \sqrt{(R^2 + X_L{}^2)}$
9. $X_C = \dfrac{E_C}{I_C}$
11. $R = \dfrac{P}{I^2}$
13. $F = \dfrac{1}{[2\pi \times X_c]}$
15. $R = \dfrac{1}{\sqrt{[(\frac{1}{Z})^2 - ((\frac{1}{X_L})^2 - (\frac{1}{X_C})^2)]}}$

Exercise 6-4

1. $E = 60$
3. $I_T = 20$
5. $I = 2$
7. $C = 11.66$
9. $R = 2.50$
11. $R_T = 6.67$
13. $I = 3.16$
15. $B = 75.99$ or 76

CHAPTER 6
Algebraic Functions

Mean and Median Value Exercise

Exercise 6-1

1. Mean: 26.3, Median: 12
3. Mean: 31.6, Median: 7
5. Mean: 496.6, Median: 54

CHAPTER **7**
Geometric Functions

Perimeter and Circumference Exercises

Exercise 7-1

1. 10'-6"
3. 12'-3"
5. 18.5' or 18'-6"
7. 120'

Exercise 7-2

1. Circumference: 78.50", Diameter: 25", Radius: 12.5"
3. Circumference: 8', Diameter: 2.54', Radius: 1.27'

Polygon Area Exercise

Exercise 7-3

1. 161.25 ft^2 or 23,220 in^2
3. 90.86 ft^2 or 13,083 in^2
5. 53.395 ft^2 or 7,688.96 in^2 or 4.96 m^2
7. 84,430.50 in^2 or 586.26 ft^2

Combined Area Exercises

Exercise 7-4

1. 5.96 ft^2 or 8.19 in^2
3. 600 ft^2
5. 1,650.96 m^2

Practical Exercise 7-5

1. 440 ft^2
3. 718 ft^2 [(30' × 17') + (22' × 8') + (0.5 × 8' × 8') = 718 ft^2]
5. 4,000 watts [20' × 25' = 500 ft^2 × 8 = 4,000 watts]
7. 5,154.5 VA [1.3 × 3,965]

Cubic Volume Exercise

Exercise 7-6

1. 1,296 in^3 or 0.75 ft^3
3. 1,023.75 m^3

Cylindrical Volume Exercises

Exercise 7-7

1. 8,010.4 ft^3 [24'-6" (24.5') ÷ 2 = 12.25' × 12.25' = 150.0625 ft^2 × 3.14 = 471.2 ft^2 (rounded from 471.19625 ft^2) × 17' = 8,010.4 ft^3]
3. 2,171.07 ft^3 [9'-3" (9.25') ÷ 2 = 4.63' (rounded from 4.625') × 4.63' = 21.44 ft^2 (rounded from 21.4369 ft^2) × 3.14 = 67.32 ft^2 (rounded from 67.3216 ft^2) × 32.25' = 2,171.07 ft^3]

Practical Exercise 7-8

1. 14.08 ft^3 [(1.5/2 or 0.75)^2 × π = 1.76 (rounded from 1.7584) × 8 = 14.08 ft^3]
3. 10.43 yd^3 [281.60 ft^3 ÷ 27 = 10.43 yd^3 (rounded from 10.4296)]

CHAPTER **8**
Trigonometric Functions

Side and Angle Exercises

Answers will vary depending on if students use a scientific calculator or the trigonometric chart.

Exercise 8-1

1. 42° or 41.8°
3. 180°
5. 75° or 74.8°
7. 63.6°
9. 33° or 32.6°

Exercise 8-2

1. 22.47 or 22.45
3. 61.18 or 61.32
5. 26.38 or 26.41
7. 169.17 or 167.82
9. 14.36 or 14.43

Exercise 8-3

1. 32.31
3. 21.66
5. 12
7. 70.71
9. 30.85

Practical Exercise 8-4

1. 14.1
3. 10.3 or 10.5
5. 6.7

Practical Exercise 8-5

1. 128.06
3. 244.95

CHAPTER 9
Electrical Measurement and Calculation

Electrical Measurement Exercise

Exercise 9-1

1. Clamp-on meter
3. 600 V
5. 400 A
7. 400 A

Ohm's Law Exercise

Exercise 9-2

1. $R = 4\ \Omega$
3. $E = 115.05$ V
5. $E = 230$ V
7. $R = 24\ \Omega$

Power Formula Exercises

Exercise 9-3

1. $P = 3,600$ W
3. $E = 120$ V
5. $E = 120$ V
7. $P = 4,320$ W

Practical Exercise 9-4

1. $I = 6.25$ A
3. $I = 12.8\ \Omega$
5. $I = 6$ A

Power Factor Exercises

Exercise 9-5

1. 92.31%
3. 98.91%

Exercise 9-6

1. 3,026.4 W
3. 9,408 W
5. 21,374.5 W
7. 63,110.4 W

Series Circuit Exercise

Exercise 9-7

1. $R_2 = 3.8$, $R_T = 12$, $P = 1,200$
3. $I = 8.35$, $R_T = 28.75$, $P = 2,004$

Parallel Circuit Exercise

Exercise 9-8

1. Total current = 5.5, Total resistance = 21.82, $P = 660$
3. Total current = 10, Total resistance = 24, $P = 2,400$

Series-Parallel Circuit Exercise

Exercise 9-9

1. $R_T = 11.67\ \Omega$
3. $R_T = 387.5\ \Omega$

CHAPTER 10
Practical Applications

Practical Exercise 10-1

1. A. 1
   B. 1
   C. 1
   D. 2
   E. 5
   F. 3
   G. 1
   H. 2
   I. 1
   J. 2
   K. 19

3. 2

5. A. 1
   B. 1
   C. 1
   D. 0
   E. 1
   F. 1
   G. 1
   H. 1
   I. 1
   J. 1
   K. 9

Practical Exercise 10-2

1. 323′

3. 266′

5. 589′

7. 456′

Practical Exercise 10-3

1. 1

3. $280

5. $1,716

7. $360

9. A. $1,767
   B. $360
   C. $438
   D. $1,716
   E. $95
   F. $280
   G. $420
   H. $440
   I. $5,516

Practical Exercise 10-4

1. $22,197.96 [$22,536 − ($22,536 × 0.015)]

3. $21,679.20 [$408 ÷ 1,000 ft = $0.408 × 300 ft = $122.40 × 3 = $367.20 (three color wires) + $10,248 (light poles) + $11,064 (lights) = $21,679.20]

5. $180
   25 minutes × 2 (each way) = 50 minutes
   35 minutes × 2 (each way) = 70 minutes
   50 minutes + 70 minutes = 120 minutes
   120 minutes ÷ 60 minutes/hour = 2 hours
   2 hours × 5 days = 10 hours
   10 hours × $18 = $180

7. $276.48
   $7.54 × 12 = $90.48
   $11.30 × 12 = $135.60
   $0.60 × 84 = $50.40
   $90.48 + $135.60 + $50.40 = $276.48

9. $10,224.32 [$35,200 − $24,975.68 = $10,224.32]

Practical Exercise 10-5

1. 1,750 ft^2

3. 1,407.13 ft^2
   Area of living room, dining room, study, kitchen, part of hall, and part of laundry room: 33′ × 35′ = 1,155 ft^2
   Area of other part of laundry room, other part of hall, and bathroom: 7′ × 28′ = 196 ft^2
   Area of cutout on front of dining room: 12′-3/4″ × 2′ = 24.125 ft^2
   Area of cutout on side of dining room: 8′ × 2′ = 16 ft^2
   Area of cutout on side of kitchen: 8′ × 2′ = 16 ft^2
   1,155 ft^2 + 196 ft^2 + 24.125 ft^2 + 16 ft^2 + 16 ft^2 = 1,407.13 ft^2 (rounded from 1,407.125 ft^2)

5. 12,664.17 VA [4,221.39 VA × 3]

7. 23 bags [10 ÷ 0.45]

9. 5.58 yd^3 [3.14 × 6 = 18.84 × 8 = 150.72 ft^3 ÷ 27 = 5.58 yd^3]

11. 24.70 yd^3
    Duct bank volume: 2′ × 3′ × 120′ = 720 ft^3
    PVC raceways volume: 3.14 × 2.25^2 × 1,440 (120 × 12) × 4 = 91,562.4 in^3 × 0.0005787037 = 52.99 ft^3
    720 ft^3 − 52.99 ft^3 = 667.01 ft^3 ÷ 27 = 24.70 yd^3

Practical Exercise 10-6

1. $80,725.25 [(240 fixtures × 458 watts × 17 hours/day × 360 days × $0.12 kWh) ÷ 1,000 = $80,725.25]

3. $54,286.85 [$80,725.25 − $26,438.40 = $54,286.85]

5. 1.66 years or about 1 year and 8 months [$90,038.40 ÷ $54,286.85]

7. $86,645.25 [$5,920 + $80,725.25 (yearly utility cost)]

Practical Exercise 10-7

1. 6.97′

3. 15.56″

5. 8″

7. 54.08 Ω

9. 1,322.86 W

Practical Exercise 10-8

1. 93 Ω

3. 15 Ω

5. 40 V

7. 12.5 Ω

9. 5 A

Practical Exercise 10-9

1. A. E_1: 20 V
   B. E_2: 40 V
   C. E_3: 50 V
   D. E_4: 10 V
   E. I_1: 2 A
   F. I_2: 2 A
   G. I_3: 2 A
   H. I_4: 2 A
   I. I_T: 2 A
   J. R_T: 60 Ω
   K. P_1: 40 W
   L. P_2: 80 W
   M. P_3: 100 W
   N. P_4: 20 W
   O. P_T: 240 W
3. A. E_1: 120 V
   B. E_2: 120 V
   C. E_3: 120 V
   D. I_1: 6 A
   E. I_2: 4 A
   F. I_3: 2 A
   G. I_T: 12 A
   H. R_T: 10 Ω
   I. P_1: 720 W
   J. P_2: 480 W
   K. P_3: 240 W
   L. P_T: 1,440 W
5. A. E_1: 93.75 V
   B. E_2: 56.25 V
   C. E_3: 56.25 V
   D. I_1: 3 A
   E. I_2: 2.25 A
   F. I_3: 0.75 A
   G. I_T: 3 A
   H. R_T: 50 Ω
   I. P_1: 281.25 W
   J. P_2: 126.56 W
   K. P_3: 42.19 W
   L. P_T: 450 W

Practical Exercise 10-10

1. 46.15 A
3. 15 A
5. 4.32
7. 11.88 Ω
9. 13 receptacles [20 × 120 = 2,400 ÷ 180 = 13.33 (rounded to 13)]